YOUR ULTIMATE TRAVEL ADVENTURE LIST

Commissioned and project managed by Duck Egg Blue Limited

Author: Anna Brett
Publishing Director: Piers Pickard
Publisher: Rebecca Hunt
Editorial Director: Joe Fullman
Art Director: Andy Mansfield
Print Production: Nigel Longuet

Published in October 2024
by Lonely Planet Global Limited
CRN: 554153
ISBN: 978-1-83758-306-5
1 3 5 7 6 9 10 8 6 4 2

Printed in Malaysia

Stay in Touch
Lonelyplanet.com/contact

Lonely Planet Office:
IRELAND
Digital Depot, Roe Lane (off Thomas St), Digital Hub, Dublin 8, D08 TCV4, Ireland

FSC www.fsc.org MIX Paper | Supporting responsible forestry FSC™ C021741

Paper in this book is certified against the Forest Stewardship Council™ standards. FSC™ promotes environmentally responsible, socially beneficial and economically viable management of the world's forests.

lonely planet KIDS

YOUR ULTIMATE TRAVEL ADVENTURE LIST

ANNA BRETT

CONTENTS

Colorful hot-air balloons drift over Cappadocia in Turkey (see p.108)

YOUR ULTIMATE TRAVEL ADVENTURE AWAITS

Get ready for skywalks, roller coasters, and ice-skating in North America, as well as relaxing among natural wonders like waterfalls, cloud forests, and hot springs. Don't forget there are big-city sights and even bigger outer-space quests here too.

If you could jump on a plane tomorrow and go anywhere in the world, where would you choose? There's so much to see out there, it's hard to know where to begin! You'll probably want to see the famous sights, have fun doing activities you can't do at home, and meet unusual animals you've only read about in books. But what about also trying things you've never even heard of, or challenging yourself to an adrenaline-filled adventure? If you need inspiration for creating a list of the best experiences on the planet and the most exciting ways to try them, this book has all the answers!

You can trek, horseback ride, and kayak your way to some of the experiences in South America, but also enjoy staying in one place for a few hours at a soccer match or in a mud volcano! Then there's the Amazon Rainforest — which is way more than just one experience.

Northern European experiences include chasing the Northern Lights, sleeping in an ice hotel, and paragliding over a castle. Meanwhile, in Central and Southern Europe, you can come face-to-face with volcanoes, vampires, and... tomatoes!

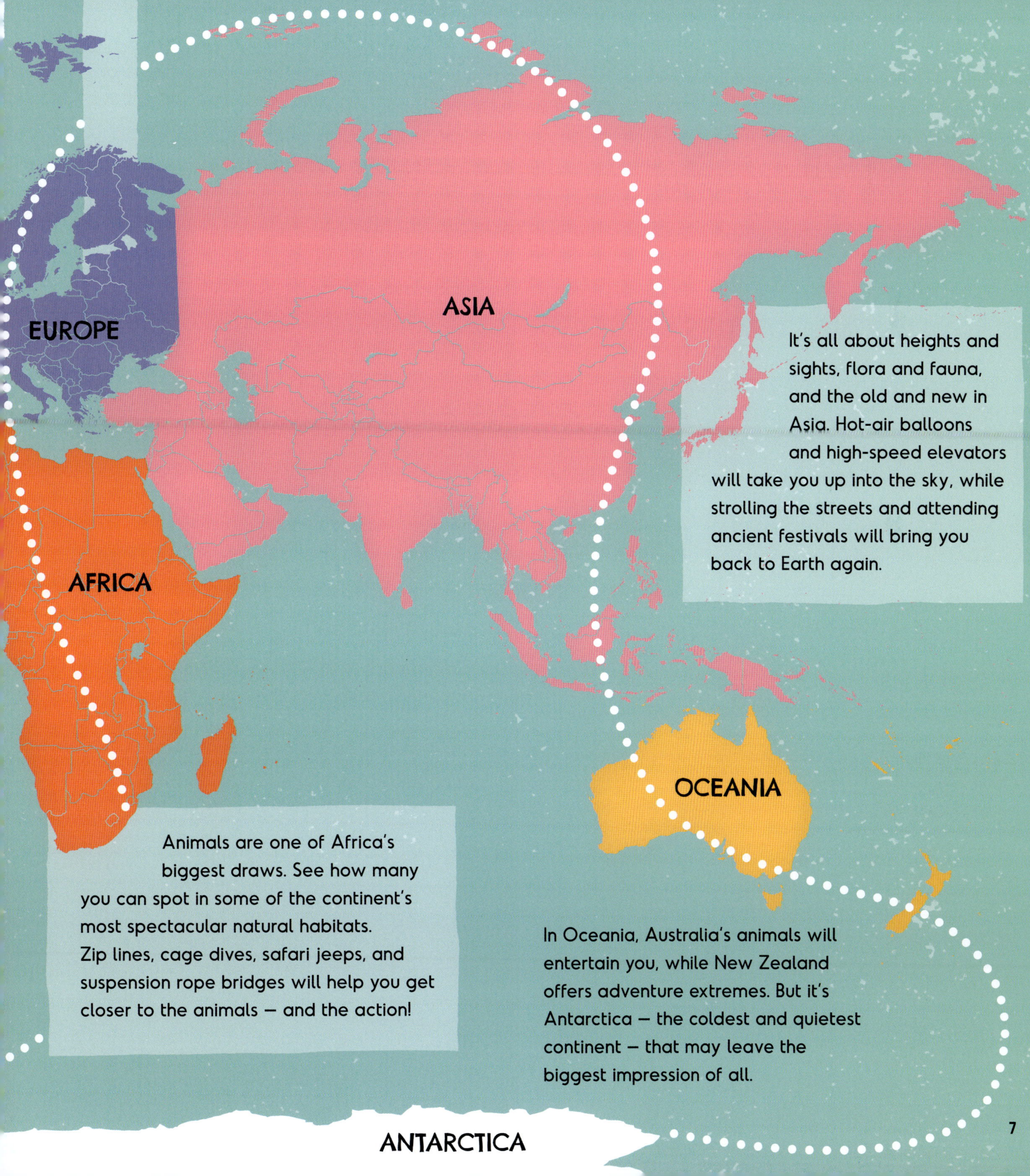

It's all about heights and sights, flora and fauna, and the old and new in Asia. Hot-air balloons and high-speed elevators will take you up into the sky, while strolling the streets and attending ancient festivals will bring you back to Earth again.

Animals are one of Africa's biggest draws. See how many you can spot in some of the continent's most spectacular natural habitats. Zip lines, cage dives, safari jeeps, and suspension rope bridges will help you get closer to the animals – and the action!

In Oceania, Australia's animals will entertain you, while New Zealand offers adventure extremes. But it's Antarctica – the coldest and quietest continent – that may leave the biggest impression of all.

YOUR TYPE OF TRIP

Staring at a map of the world trying to figure out where you want to go can be a little overwhelming. Sometimes the best way to plan a trip is to decide what sort of thing you want to do. So, do you want to see amazing animals, take part in adrenalin-fueled activities, go on rides through unknown lands, visit bustling cities, stare wide-eyed at incredible buildings and views, get hands on at museums, join the crowds for incredible experiences – or all of the these? The suggestions on these pages should help you choose the kind of travel adventure you want to go on.

6 ICONIC LANDMARKS

- Statue of Liberty, USA, p.14
- Christ the Redeemer, Brazil, p.43
- Tower of London, UK, p.72
- Great Pyramids, Egypt, p.88
- Great Wall of China, p.120
- Sydney Opera House, Australia, p.138

8 AMAZING ACTIVITIES

- Skating on Lake Louise, Canada, p.28
- Horseback riding in Chile, p.54
- Paragliding over a castle in Germany, p.84
- Zip-lining in South Africa, p.104
- Surfing on the Gold Coast, Australia, p.136
- Snorkeling in the Great Barrier Reef, Australia, p.134
- Bungee-jumping in New Zealand, p.148
- Kayaking to see penguins in Antarctica, p.154

6 ASTOUNDING ANIMALS

- Giant tortoises on the Galápagos Islands, Ecuador, p.40
- Great white sharks in South Africa, p.100
- Tigers in Ranthambore National Park, India, p.114
- Snow monkeys in Japan, p.122
- Crocodiles on the Adelaide River, Australia, p.132
- Poison dart frogs in the Amazon Rainforest, Brazil, p.44

5 GREAT WAYS TO GET WET

- Walking beneath Niagara Falls, Canada–USA, p.26
- Diving to the Grenada Underwater Sculpture Park, p.34
- Staring up at Angel Falls, Venezuela, p.52
- Taking a helicopter ride over Victoria Falls, Zambia–Zimbabwe, p. 102
- Whale-watching, Australia, p.140

8 THINGS TO DO IN THE DARK

- Stargazing in the Atacama Desert, Chile, p.56
- Watching the Northern Lights, Norway, p.62
- Touring the Catacombs of Paris, France, p.76
- Exploring the Mount Etna caves, Italy, p.83
- Playing in Gardens by the Bay, Singapore, p.126
- Watching fireworks in Sydney, Australia, p.139
- Touring the Derinkuyu Underground City, Turkey, p.109
- Looking at glowworms in the Waitomo Caves, New Zealand, p.144

6 INCREDIBLE RIDES

- Kingda Ka roller coaster, USA, p.20
- Dog sled, Norway, p.62
- River Nile cruise boat, Egypt, p.89
- Jet boat, Australia, p.141
- Icebreaker, Antarctica, p.154
- Bullet train, Japan, p.125

10 SPECIAL CITIES

- New York, USA, p.14
- Rio de Janeiro, Brazil, p.42
- Buenos Aires, Argentina, p.48
- London, UK, pp.70–73
- Paris, France, p.76
- Marrakesh, Morocco, p.92
- Cape Town, South Africa, p.101
- Beijing, China, p.120
- Tokyo, Japan, p.124
- Sydney, Australia, p.138

6 STUNNING VIEWS FROM ABOVE

- Grand Canyon Skywalk, USA, p.12
- Machu Picchu, Peru, p.46
- Eiffel Tower, France, p.77
- Table Mountain cable car, South Africa, p.101
- Burj Khalifa, UAE, p.110
- Fairy chimneys of Cappadocia by hot-air balloon, Turkey, p.108

6 MONUMENTAL MOUNTAINS, VOLCANOES AND GEYSERS

- Yellowstone National Park, USA, p.18
- Totumo Mud Volcano, Colombia, p.50
- Mount Stromboli & Mount Etna, Italy, p.82
- Dallol Hydrothermal Field, Ethiopia, p.94
- Mount Everest, Nepal–China, p.116
- Rotorua, New Zealand, p.146

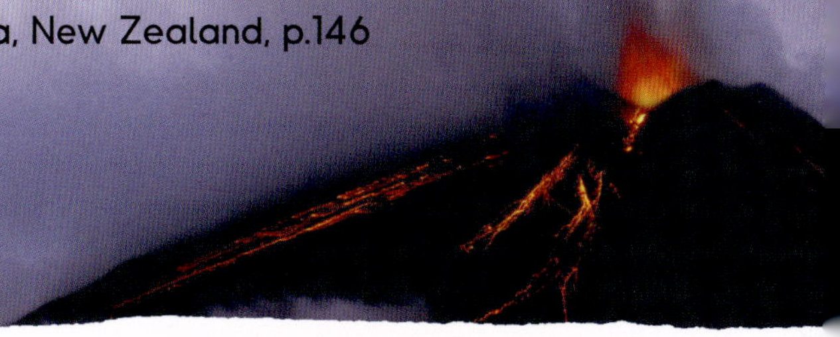

6 SUPER EXPERIENCES

- Watch a rocket launch at the Kennedy Space Center, USA, p.24
- Dance at the Rio Carnival, Brazil, p.42
- Stay at an ice hotel, Sweden, p.66
- Get messy at La Tomatina, Spain, p.80
- Light up the sky at the Lantern Festival, Vietnam, p.128
- Visit a research station, Antarctica, p.156

8 FUN-PACKED MUSEUMS

- Hollywood Museum, USA, p.16
- Royal Tyrrell Museum, Canada, p.30
- V&A Museum, UK, p.71
- Louvre Museum, France, p.77
- Universe, Denmark, p.69
- Grand Egyptian Museum, Egypt, p.89
- Terracotta Army Museum, China, p.118
- National Museum of Nature and Science, Japan, p.124

Let's start with...

...AWESOME

NORTH AMERICA

Start your world tour in North America for a taste of every type of experience going. Your heart will race as you zip down the world's tallest roller coaster and take on the rides at numerous theme parks, but you'll also find calm in the natural beauty of the world's most beautiful outdoor ice rink and a unique cloud forest. There are iconic city sights to see in New York and fascinating festivals in Mexico, so prepare to blast into your ultimate travel adventure with this continent of varied experiences. And when we say blast, there are also a few rocket launches you can watch too!

PLACES TO VISIT:

1. **Royal Tyrell Museum & Badlands**, Alberta, Canada, p.30
2. **Lake Louise**, Banff National Park, Alberta, Canada, p.28
3. **Yellowstone National Park**, Wyoming, USA, p.18
4. **Niagara Falls**, Canada–USA border, p.26
5. **New York**, New York, USA, p.14
6. **Six Flags Amusement Park**, New Jersey, USA, p.20
7. **Grand Canyon**, Arizona, USA, p.12
8. **Hollywood**, Los Angeles, California, USA, p.16
9. **Orlando**, Florida, USA, p.22
10. **Kennedy Space Center**, Florida, USA, p.24
11. **Everglades**, Florida, USA, p.23
12. **Mexico City**, Mexico, p.32
13. **Teotihuacán**, Mexico, p.33
14. **Underwater Sculpture Park**, Grenada,, p.34
15. **Monteverde Cloud Forest**, Costa Rica, p.36

NORTH TO SOUTH

The continent of North America stretches from Canada in the north to Panama in the south and includes the islands of the Caribbean too. It's the world's third-largest continent and its landscape is as varied as the cultures of the 23 countries within it.

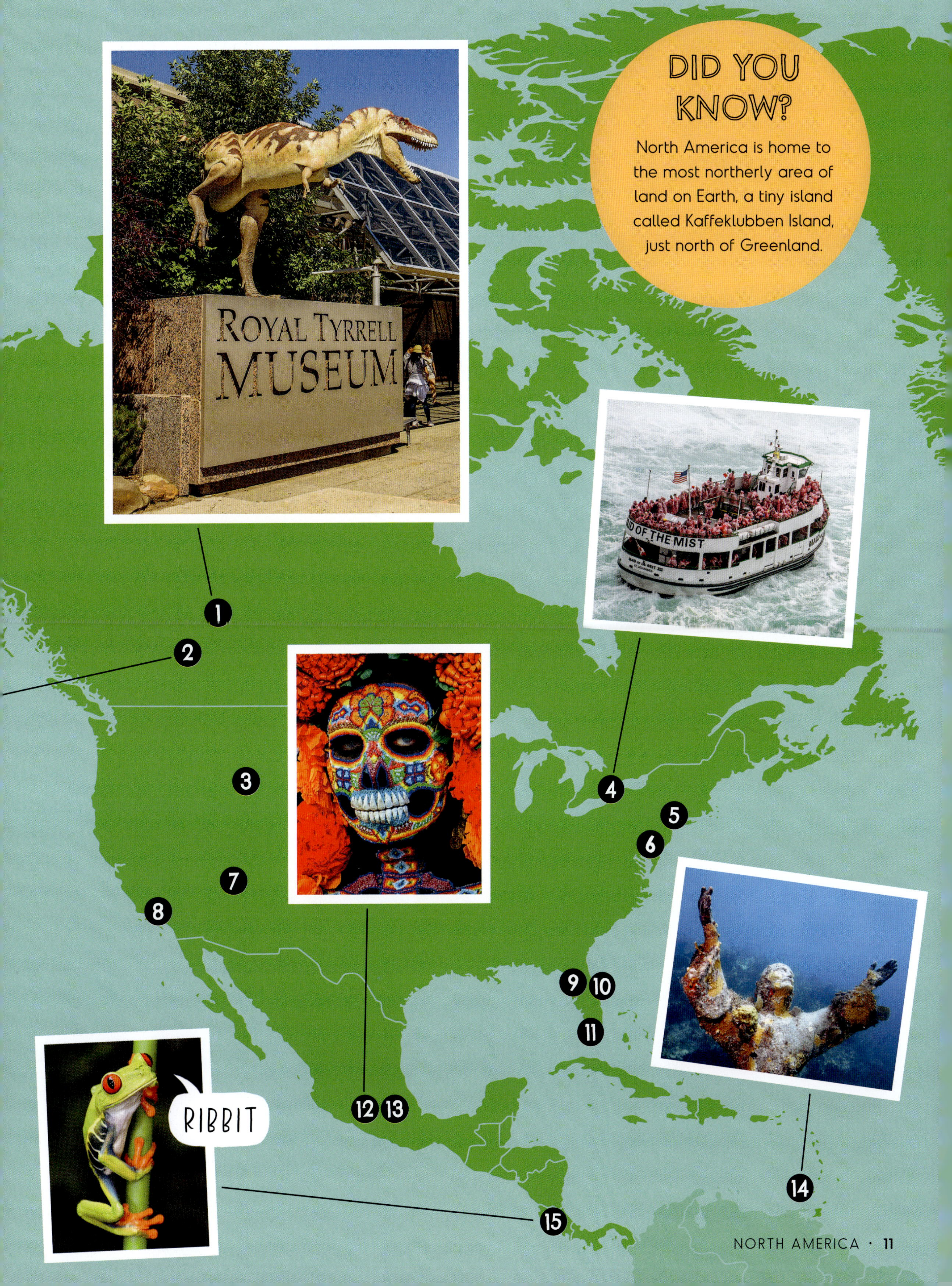

DID YOU KNOW?

North America is home to the most northerly area of land on Earth, a tiny island called Kaffeklubben Island, just north of Greenland.

Stare down into our planet's history...

...ON THE GRAND CANYON SKYWALK

Would you be brave enough to step out onto a glass bridge over 4,000 ft. (1,000 m) up in the air? And then look directly down into the depths of a canyon that's nearly 6 million years old? The Grand Canyon Skywalk offers you that chance – an experience that may seem scary at first, but which is perfectly safe! The Grand Canyon is a natural rock formation in Arizona, USA, and the glass skywalk is a thrilling way to see it.

RIVER ROUTE

The canyon is a narrow valley that cuts between layers of rock to show nearly 2 billion years of planet Earth's history. The skywalk is the best way to see right down into the bottom of the canyon to the Colorado River. It was this river that started cutting a channel through the rock millions of years ago. As it eroded the rock, it exposed many different layers. In places, you can see fossils of ancient marine animals that date back 1.2 billion years!

DID YOU KNOW?

The skywalk is strong enough to hold 70 fully loaded airplanes!

THE STATS

Location: Arizona, USA
Best time to visit: Mar.–May and Sept.–Nov.
Height: 4,770 ft. (1,450 m)
Floor: Five layers thick
Opened: 2007

WORLD WONDER

Step out onto the glass bridge and you'll instantly understand why the Grand Canyon is one of the Seven Natural Wonders of the World. And you'll be forgiven if your legs start to shake. Walking out into the sky may seem nerve-racking at first – the canyon is 6,070 ft. (1,850 m) deep in places!

MORE FUN!

There are lots of other ways to explore and experience the canyon. Why not try some of the following?

Take a helicopter flight to see some of its amazing features from the air, such as Horseshoe Bend (right), a large meander in the Colorado River.

Hike or ride a mule (below) down to the bottom of the canyon. But be warned, it can take 4–5 hours to trek all that way – and double that coming back up again.

Take a rafting trip along the river (right). The view will be a little different down here, but just look up to remind yourself where you came from!

Take a bite and savor the sights...

...OF THE BIG APPLE

New York is the city that charms everyone who visits it. It's known as the "Big Apple," though no one is exactly sure why. But what is certain is that if you pick New York as your destination, you're picking the best fruit on the tree! It's described as "the city that never sleeps," and there are endless options for culture, entertainment, and food. So spend some time in town and see how much you can bite off!

THE STATS

Location: New York, USA
Best time to visit: Apr.–May & Sept.–Oct.
Population: 8.3 million
Languages spoken: Over 800
Height of Statue of Liberty: 305 ft. (93 m)
Height of Empire State Building: 1,455 ft. (450 m), including the spire and tip

START WITH THE STATUE

Begin your day at the Statue of Liberty. This iconic monument stands tall on Liberty Island and represents freedom, democracy, and friendship. Given as a gift to America by France in 1885, the figure represents Libertas, the Roman goddess of freedom. Visiting "Lady Liberty" is a must-do New York experience, and climbing at least 162 steps to reach the crown is an extra achievement. The crown's seven spikes represent the seven oceans and seven continents of the world.

BIRD'S-EYE VIEW

There's so much to see in NYC, one of the best ways to figure out what to do is to head up high. Get a bird's-eye view of the city from New York's most famous building, the Empire State Building. Built in 1930, it is 102 stories high. Head out to the open-air observatory on the 86th floor and take in 360-degree views. Or head to the 102nd floor – you'll be able to see as far as 80 miles (130 km) on a clear day. The Empire State Building was built in a record-breaking 1 year and 45 days – just in time for King Kong to climb it.

OUTDOOR PLAY

The island of Manhattan is the largest of New York's five boroughs. It's here that skyscrapers rise high, lights shine bright, and people flood in for work and fun every single day. In the middle of the island you'll find Central Park, a large green space that offers everything an urban park should – beautiful lakes, landscaped terraces, fantastic sculptures, meandering pathways, and huge concert areas. It even has 21 different playgrounds to choose from!

GETTING AROUND

One way to zip across town is in one of 13,587 yellow cabs. These are the only vehicles in the city allowed to pick up passers-by. To "hail" a cab, just make yourself known in whichever way you please – wave, whistle, or simply walk up and get in! Or if you need to venture off the island of Manhattan to another district, cross the water on the NYC Ferry. Queens is famous for its diversity, Brooklyn for being trendy, the Bronx is the home of hip-hop music, and Staten Island offers acres of green space.

BRILLIANT BROADWAY

As darkness falls, the lights shine on Broadway and its 41 theaters. This iconic street is the home of the performing arts. Some of the world's most skilled actors, singers, dancers, and musicians work here. It's not just the stars on stage who are brimming with talent – you'll find plenty of creativity behind the scenes, with award-winning musical scores, carefully hand-stitched costumes, and spectacular set designs on display.

DID YOU KNOW?

The first pizzeria in the USA opened in New York in 1905.

Experience movie magic...

...IN HOLLYWOOD

What could be more exciting than hanging out with the stars? Head to California and turn your dreams into reality as you visit the home of America's film and television industry – Hollywood! Take a studio tour to step behind the scenes of a movie set, visit the Hollywood Museum to learn how and why this neighborhood has become famous worldwide, and then try to spot your favorite actors on the Walk of Fame. Get your glad rags on, it's time for some good old razzle-dazzle entertainment!

WOW, WARNER BROS.!

Visit the Warner Bros. Studio Tour to explore real sets and find out about more than 100 years of cinematic storytelling. You can also check out props and costumes from a variety of movies, including the DC Universe series of films (below). Warner Bros. was the first studio to produce a "talkie" film (*The Jazz Singer*, in 1927), in which the characters' dialogue could be heard. Before that all movies were silent!

UNBELIEVABLE UNIVERSAL!

Universal Studios is a working film studio, offering tours as well as a movie-based amusement park. The park is full of themed areas, the largest being the Wizarding World of Harry Potter. Meet your favorite characters and ride family-friendly roller coasters, then head off to explore the rest of the park. Catch the show Waterworld: A Live Sea War Spectacular (below), and you'll witness professional stunt-people surviving perilous plunges, fearsome fire fights, epic explosions, and even walk away from a plane crash. It's high drama!

THE HISTORY OF HOLLYWOOD

If you're a movie buff you'll be in your element at the Hollywood Museum, surrounded by over 11,000 memorabilia items. It's an impressive collection of costumes, props, scripts, and even stop-motion figures. Before it became a museum in 1994, world-famous movie stars such as Marilyn Monroe would come to this very building to get their hair and make-up done by the beautician and cosmetic king, Max Factor.

WALK OF FAME

Take a stroll along Hollywood Boulevard and you'll pass more than 2,700 stars embedded in the ground. The stars show the names of actors, directors, musicians, and producers... all people who have had great achievements in the entertainment industry. Emblems on each star indicate which of six categories the person is being recognized for: motion pictures, TV, music, broadcast radio, theater, or sports entertainment. One person has stars in five categories – Gene Autry, a celebrity famous during the 1930s–1950s.

LETTERS ON THE HILLSIDE

Look up and you'll know you're in Hollywood... a large sign spells it out for you! These nine letters have become a cultural icon, even though they were only ever intended to be a temporary advertisement for a new housing development when they were first erected in 1923.

THE STATS

Location: Los Angeles, California, USA
Best time to visit: Nov.–Feb. for award season
First Hollywood studio: Nestor Studios in 1911
First feature length film: *The Squaw Man* in 1914
First Academy Awards ("Oscars") held: 1929

MORE FUN!

Of course, there's no better place to watch a movie than Hollywood. LA is home to some of the most outstanding movie theaters in the country, from the historical to the state of the art. Grauman's Egyptian Theatre (right) opened in 1922, just before the real King Tutankhamun's tomb was found. The Regal LA Live is one of the largest cinemas in the country and has a 4DX auditorium – it's a theater and theme park ride rolled into one! Or if you have cash to splash, you could even buy yourself a ticket to a red carpet film premiere event.

Discover a rainbow on the ground...

...IN YELLOWSTONE NATIONAL PARK

Get ready to explore Yellowstone and discover why it's different from the other 62 national parks in America – it sits on top of a dormant volcano! This underground volcanic system heats water on the Earth's surface, creating hydrothermal features such as geysers and mud pools. There are over 10,000 features to encounter here, so it's no wonder this area is an explorer's delight.

THE STATS

Location: Wyoming, as well as Montana and Idaho, USA
Best time to visit: Apr.–Sept.
Established: 1872
Size: Around 3,400 sq. miles (8,800 sq km)
Old Faithful eruption height: 100–180 ft. (30–55 m)

HOT PROPERTY

There are five different types of hydrothermal features to be seen in Yellowstone. Check them off your list as you walk around.

Geysers – these erupt boiling water and steam into the sky.

Hot springs – pools of heated water.

Fumaroles – holes in the ground that shoot out steam.

Travertine terraces – steps naturally formed from minerals deposited by the hot springs.

Mud pots – hot springs that dissolve the rock around them, creating bubbling mud pools.

CHASE THE RAINBOW

The Grand Prismatic Spring is the main focus for many Yellowstone visitors. It's the largest (and most colorful) hot spring in America. Take a moment to appreciate this rainbow pool and you'll understand how it got its name – *Prismatic* means "brilliantly colored." It's surrounded by bands of brilliant colors, created by bacteria called thermophiles (meaning "heat lovers"). Some species of thermophiles actually prefer the cooler water around the edge of the pool, but others prefer the deeper water, which is much warmer. There are no vivid colors in the middle though – the center of the spring is too hot for any bacteria to live in.

OTHER POOLS AND GEYSERS

The water reaches 160°F (70°C) in the Grand Prismatic Spring, so you must stay on a boardwalk for safety. As you follow the path, you'll also pass the Turquoise Pool, Opal Pool, and Excelsior Geyser. The best time to visit is around lunchtime on a sunny day – earlier in the day, the spring is often covered in mist due to the cooler air temperature above it.

DID YOU KNOW?

Yellowstone has more hot springs and geysers than any other place in the world.

OLD FAITHFUL

A visit to Yellowstone isn't complete without trying to predict when Old Faithful geyser will erupt. A geyser occurs when water in a channel underground is heated under high pressure, resulting in an eruption through a hole in the surface. This particular jet of water and steam tends to shoot out around every 90 minutes or so. See if you can guess the exact minute!

WATCH OUT FOR WILDLIFE

Wildlife loves Yellowstone! More than 200 species of animals call the park home, and tourists love to come and spot grizzly bears (right) and bald eagles, as well as the geysers and hot springs. However, it's best to bring bear spray with you – these beasts ae not the cuddly kind.

Try not to scream...

...ON THE WORLD'S TALLEST ROLLER COASTER

Some experiences on your travels around the world will make your eyes sparkle or your toes twitch, but none will make your jaw drop quite like this one. If you want to experience a true adrenaline rush, then head for the highest height and zoom down it on a roller coaster called Kingda Ka! It is the tallest 'coaster in the world, and the fastest in North America, so a true once-in-a-lifetime experience (unless you dare to ride it more than once!).

TOP TRACK

Kingda Ka can be found at the Six Flags Great Adventure amusement park in Jackson, New Jersey, USA. It's been thrilling people since 2005 and cost US$25 million to make. Each train can carry 18 passengers, and there are four trains that shoot their way up and over an upside-down, U-shaped track. The train's bright colors are easy to spot on the lime-green track when stationary, but it all becomes a blur when racing at such high speed.

ROCKETING UP

In order to reach such a height, Kingda Ka's train needs a rocket-like launch. This specially designed one boosts the train you're riding in from zero to top speed in just 3.5 seconds.

DID YOU KNOW?

There are over 2,400 roller coasters around the world.

TWISTING DOWN

Once at the top of the track, you may get half a second to catch your breath... before the descent begins and you drop down and twist in a 270-degree spiral! To add even more excitement, there's then a second 129 ft. (39 m) hump that will lift your rear off your seat and give you a moment of weightlessness.

THE STATS

Location: Six Flags Great Adventure, New Jersey, USA
Best time to visit: May–Jun.
Height: 456 ft. (139 m)
Length: 3,118 ft. (950 m)
Duration: 50 seconds
Top speed: 128 mph (206 km/h) in 3.5 seconds

SECOND TIME LUCKY

With such a height to climb, it's understandable that sometimes the train on this roller coaster might not make it quite to the top! When this happens, it's called a rollback. It just means there wasn't enough force in the launch, and the train safely rolls back into the station to try again.

DROP OFF

If the world's tallest roller coaster leaves you wanting even more thrills, you don't need to look far. The world's tallest and fastest drop tower is attached to the tower of Kingda Ka. Zumanjaro: Drop of Doom is 415 ft. (126 m) high and drops willing passengers down at 90 mph (145 km/h). Whoosh!

MORE FUN!

Six Flags is one of the world's largest amusement parks, covering an area of over 510 acres (210 hectares). It boasts a 345-acre (140-hectare) animal park – home to lions, tigers, bears, and elephants – plus two water rides and no fewer than 14 roller coasters.

Seek thrills and adventure...

... IN ORLANDO'S

THEME PARKS

Florida is home to a whopping nine theme parks, and millions of visitors from around the real world visit these fantasy worlds every year to enjoy the entertainment (as well as the sunshine). Which park will you choose to visit?

Half theme park, half zoo, Disney's Animal Kingdom is where nature provides the entertainment. Don't miss the epic Tree of Life sculpture (above), as well as the chance to rescue a dinosaur at DinoLand USA.

WALT DISNEY WORLD

The pick of the bunch for Disney fans, Walt Disney World® is actually made up of four theme parks – and you'll find two water parks and a golf course here too! The largest theme park resort in America, Disney World calls itself "the most magical place on Earth." At Magic Kingdom, the original park that opened in 1971, visitors can discover Cinderella's Castle (main image), ride the Seven Dwarfs Mine Train. and watch the "Happily Ever After" fireworks show.

The EPCOT Center is a showcase of human achievement, featuring landmarks, traditions, and cuisines from all over the world. Here you can travel through time on Spaceship Earth, a ride housed inside EPCOT's iconic giant golf ball (below).

At Disney's Hollywood Studios, enter Toy Story Land and experience thrills such as the mega-coaster in Andy's backyard. Or head to Star Wars: Galaxy's Edge (below), where a range of futuristic rides await.

THE STATS

Walt Disney World location: Orlando, Florida, USA

Best time to visit: Dec.–May for the driest weather

Opened: 1971

Most visited area: Magic Kingdom

Hotels in park: Over 30

Extra residents: More than 100 stray cats!

UNIVERSAL ORLANDO

Much like Walt Disney World®, Universal Orlando is three parks in one! Universal Studios Florida is the main attraction, with themed experiences from *Harry Potter*, *Despicable Me*, and *The Simpsons*, to name a few.

Universal Islands of Adventure welcomes you to *Jurassic Park*-themed dinosaur attractions (below), the Amazing Adventures of Spider-Man, and Seuss Landing, which brings the wacky world of the beloved Dr. Seuss books to life.

Volcano Bay is a water park within Universal. Put your swimsuit on and enjoy water slides, a lazy river, the wave pool, and plenty of other beach attractions. Control your nerves as you take on the high-energy water thrill ride Krakatau Aqua Coaster for an extra shot of adrenaline.

DID YOU KNOW?

There's also LEGOLAND® Florida and Peppa Pig Theme Park to visit in Orlando.

MORE FUN!

There's more to Orlando than its theme parks. To the south of the city you'll find the Everglades, a vast area of natural wetlands. Take an airboat tour (right) to search for the Everglades' most famous residents – alligators. These boats have a flat bottom and a huge fan on the back to shoot them over the shallow, swampy water. They travel fast, so keep your eyes peeled for the resident wildlife, which includes snapping turtles, crocodiles, water snakes, and the elusive Florida panther.

Prepare for an amazing out-of-this-world experience...

...AT THE

KENNEDY SPACE CENTER

You're in for a memorable day out when you visit NASA's Kennedy Space Center (KSC) – a link between our world and the vast reaches of our solar system. Based at Cape Canaveral, it's a museum, a theme park, a training hub, and, of course, a launch site, all rolled into one. At KSC, you can enjoy your own thrilling journey through space – without even leaving Florida!

ICONIC HISTORY

Since it was established in 1962, many iconic NASA missions have taken off from here, including the Apollo missions, which successfully landed the first people on the Moon between 1969 and 1972. It's now the primary training center for American astronauts. Rockets are also built here in the huge 526 ft. (160 m) high Vehicle Assembly Building.

LOADS OF LAUNCHES

More rockets are being launched into space nowadays than at any other time in American history. With roughly one launch a week, there's a good chance you'll see a spaceship blast off on your visit. Viewing a launch from the KSC includes live commentary from space experts – if you can hear it over the roar of the rocket engines!

DID YOU KNOW?

Over 1 million people watched Apollo 11 launch from the Kennedy Space Center in 1969, and around 650 million people around the world watched the Moon landing on television four days later!

ASTRONAUT TRAINING

Have you got what it takes to be an astronaut? Would you be eager to walk on the Moon? Or perhaps voyage farther than we've ever gone before? To get an insight into what it takes, attend a presentation by a real astronaut. They'll share firsthand experiences about the highs and lows of life in space, give you a breakdown of the rigorous training involved, and answer any questions you might have.

MISSION MUSEUM

Since its "launch" in 1965, KSC has always offered tours of its facilities. The Visitor Complex has grown over the years and countless experiences are now available. One of the highlights is getting up close to real rockets, such as the enormous *Saturn V* – a launch vehicle that was developed for the Apollo missions. There are a further nine craft to visit in the aptly named Rocket Garden (below), including rockets from the Mercury and Gemini programs.

THE STATS

Location: Florida, USA
Best time to visit: Nov.–Apr.
Launch complexes: 2
Launchpads: 4
Annual launches: 50–60

SPACE SHUTTLE *ATLANTIS*

Head back indoors to visit another of NASA's most iconic space shuttles – *Atlantis* (below). It completed 33 missions over 30 years, transporting people, supplies, and parts to the International Space Station. Here you can see it displayed in action, with its payload doors open and mechanical arm extended.

SHUTTLE LAUNCH EXPERIENCE

In the space shuttle area of the visitor complex you'll find another extra-special attraction – the Shuttle Launch Experience. Using amazing simulation techniques, you can feel what it's like to be an actual astronaut, blasting off from Earth and ascending into orbit!

Put on your rain gear for the world's wettest day trip...

...BENEATH NIAGARA FALLS

Welcome to Niagara Falls, North America's biggest waterfall. There's no need to shower today, as this attraction will wash you clean! To truly understand the power of these falls, you're going to get closer than you could ever imagine by viewing them from the appropriately named "Hurricane Deck." So put on your poncho and get ready to experience the roar of the falls.

DID YOU KNOW?

The amount of water tumbling over the falls every minute is equivalent to around a million full bathtubs!

THREE IN ONE

Niagara Falls is actually a group of three waterfalls: American Falls, Bridal Veil Falls, and Horseshoe Falls. The largest, Horseshoe Falls, straddles the border of the USA and Canada. The other two falls are located entirely on the US side of the border. Bridal Veil Falls is the smallest and is separated from the others by Luna Island and Goat Island.

HURRICANE DECK

Take an elevator 175 ft. (53 m) down to the bottom of the Bridal Veil Falls section of Niagara. Prepare your senses for an assault as the tumbling waters gush down above you just a few yards away. The noise, the spray, the rainbows in the mist – it's a jaw-dropping experience. Keep walking along the wooden pathway, and before you know it you'll be on the Hurricane Deck, just 20 ft. (6 m) from the flow.

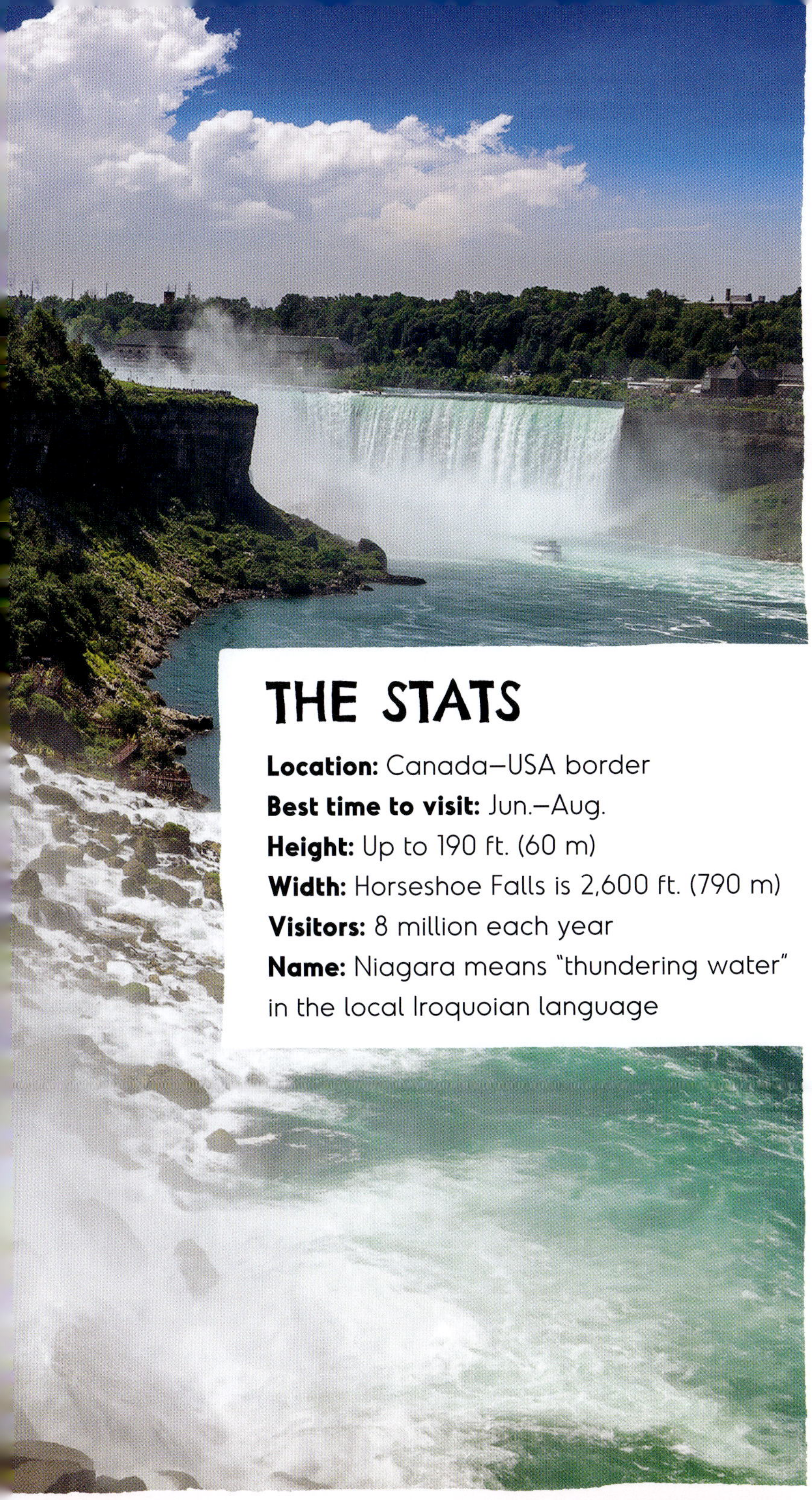

THE STATS

Location: Canada–USA border
Best time to visit: Jun.–Aug.
Height: Up to 190 ft. (60 m)
Width: Horseshoe Falls is 2,600 ft. (790 m)
Visitors: 8 million each year
Name: Niagara means "thundering water" in the local Iroquoian language

ILLUMINATIONS

As dusk descends, the falls take on a new magical identity as they're lit up by hundreds of LEDs. Special occasions are also marked by themed light shows. And, as if that wasn't enough, spectacular fireworks displays are put on nightly throughout the summer and over the Christmas period.

WALKING OVER WATER

Today, the Hurricane Deck is the closest you can get to the water rushing over these falls. Stunts and swimming are definitely not allowed! But back in the 1800s, a French daredevil named Jean François Gravelet, also known as Charles Blondin, traveled across the Niagara gorge on a tightrope. It is believed he carried out this feat around 300 times in his lifetime, often performing hair-raising, perilous tricks as he crossed!

MORE FUN!

One other way to see the falls is to hop aboard the *Maid of the Mist* boat. You'll cruise past the base of the American Falls and Bridal Veil Falls and into the basin of the Horseshoe Falls. If you want to make the experience a little more extreme, then a jet boat tour on the fall's white water rapids is also available. For a more sedate option, you can view the falls from a 520 ft. (160 m) observation tower (complete with revolving restaurant at the top) called the Skylon Tower, on the Canadian side of the border.

Ice skate on the world's most beautiful rink...

...AT LAKE LOUISE

Prepare to glide across the most picturesque ice rink you've ever seen! Lake Louise, in the Canadian Rocky Mountains, offers the world's most incredible natural skating experience, entirely for free. The views are jaw-dropping, the mountain air is fresh, and the rink is available 24 hours a day – just as long as the lake remains frozen!

SKATING AROUND

You can bring your own skates or rent them at the hotel beside the lake. If you want to get a little more competitive, you can also rent a hockey stick and puck for a game of ice hockey – Canada's national winter sport.

FRAMED BY MOUNTAINS

Lake Louise sits at the bottom of various peaks situated in the largest mountain range in North America: the Rockies. The Rocky Mountains were formed over 55 million years ago! Banff is the flagship town of the Canadian Rockies. If you stay in a mountain lodge here, you can fuel up with dinner in town when you've finished skating for the day.

DID YOU KNOW?

Lake Louise was named after Princess Louise, the fourth daughter of Queen Victoria's daughter and wife of the governor general of Canada.

RINK MAINTENANCE

The rink is cleared of snowfall each day by workers from the hotel next to the lake. A special machine called a Zamboni also helps keep skating conditions perfect by regularly resurfacing the ice. It shaves off a thin layer, cleans the surface, then sprays a fresh layer of water on top of the rink that freezes nice and smooth.

ICE CARVINGS

Every year, sculptors are invited to create beautiful ice art on the frozen lake. The Ice Magic International Ice Carving Competition (right) has been running for more than 25 years, and each year has a different theme. The competition takes place over 34 hours, and each team is given 15 ice blocks with which to work their magic. When night falls, the sculptures are illuminated by spotlights. They look even more impressive like this, as you can see the intricate details in the ice against the dark night sky.

THE STATS

Location: Banff National Park, Alberta, Canada

Best time to visit: Mid-Dec. to mid-Apr. when the lake is frozen over

Lake depth: 200 ft. (61 m)

Cost: Free to skate, just rent a pair of skates if needed

MORE FUN!

Lake Louise lies in Banff National Park, known for its dense forests and abundant wildlife – watch out for bald eagles, lynx, and flying squirrels. A train known as the Rocky Mountaineer travels between Banff and Vancouver, giving passengers spectacular views of the park and mountains. Some of its carriages have glass domes offering all-around views.

LAYER UP

You'll need to wear the right layers when out skating. Lake Louise has a subarctic climate, which means temperatures average around -0.4°F (-18°C). A hat, or "toque" as the Canadians call it, is essential for stopping heat escaping from your head. Another way to keep warm is to huddle around one of the fires on the lake's shore, or grab a hot chocolate at the ice bar!

Dig for dinosaurs...

...IN THE CANADIAN BADLANDS

Although the past has been and gone, what it leaves behind can make for some epic experiences. Dinosaur hunting is one such adventure. Of course, we all know that dinosaurs are extinct, but their fossilized bones are somewhere out there just waiting to be found. And one of the world's greatest dinosaur fossil regions can be found in the Canadian Badlands in Alberta. Can you uncover a huge reptile that used to call this region home?

THE CLUE'S IN THE NAME

The "Badlands" name comes from the quality of the soil. It's soft, muddy, won't grow crops, and can't support buildings. This also means that the top layer of soft rock is easily eroded by wind and rain to reveal a new haul of fossils that have lain in wait for millions of years. Look around and you may also spot some hoodoos, or "fairy chimneys." These are hard rock pillars left behind when the soft sedimentary rock around them eroded away.

THE STATS

Location: Alberta, Canada
Best time to visit: Mar.–May
Museum opened: 1985
Visitors per year: 400,000
Total fossils in the collection: 160,000

MEGA MUSEUM

The heart of the Badlands, and the home of everything prehistoric in this region, is the Royal Tyrrell Museum. Exploring its galleries will take you on a journey through 3.9 billion years of history. But as well as looking at what's already been found, you can also join a tour to see what is still to be discovered. Head out with a guide for a 0.9 mile (1.4 km) walk into the Badlands and touch real fossils, make a fossil cast, and watch how a dino skeleton is reconstructed.

DIG TO DISCOVER

In 1910, an American palaeontologist called Barnum Brown uncovered an impressive collection of bones from a large group of carnivorous *Albertosaurus*. These huge 30 ft. (9 m) long beasts were named after the region in which they were found – Alberta. The museum runs its own digs, so maybe you'll recreate Brown's famous discovery and find the next new dinosaur. Over 1,000 complete skeletons have been found here already, so you never know! You just need the right tools for the job – trowels, brushes, and utility knives. Looking for bones isn't always easy though. Pebbles, stones, and rocks can all be easily mistaken for fossil fragments.

DID YOU KNOW?

Removing fossils is illegal – remember to give anything you find to your guide to document and preserve.

DINOSAUR HALL

Enter the Dinosaur Hall in the Royal Tyrrell Museum and stand next to a real (fossilized!) life-size dinosaur. Look out for *Triceratops*, *Tyrannosaurus* (below), and *Camarasaurus* amongst others. This hall houses the world's largest display of dinosaur remains.

MORE FUN!

The Royal Tyrrell Museum celebrates the wonder of discovery, and what better way to get up close to the amazing finds on display than to camp next to them? For a few nights each year you can visit the museum for a sleepover – bring your sleeping bag, toothbrush, and pillow and pitch up next to the fossils. Maybe a dinosaur will even enter your dreams!

Get dressed up for...

...THE DAY OF THE DEAD FESTIVAL

Gather your family and friends and get ready for a festival like no other! This traditional Mexican celebration is a joyful event for remembering those who have passed away. On November 1 and 2 every year, Mexicans dress up, cook their favorite foods, and tell stories about their ancestors. The idea is that the festivities will encourage the souls of the deceased to visit and hear the prayers dedicated to them by their family.

CALAVERA CHARACTERS

The symbol most associated with this festival is the skull, or *calavera* in Spanish. It all started when the Mexican artist Diego Rivera (1886–1957) created a humorous artwork to remind people that underneath we are all the same: a simple skeleton. People like to paint their faces and wear elaborate suits and dresses to mimic Rivera's Calavera Catrina character. Sugar skulls are given as treats to both the living and the dead, and skeletons are posed doing funny things wherever the festival is being celebrated.

FESTIVE FOOD

If you're a spirit returning to the world of the living for the festival, it's likely you'll be hungry and thirsty after all that travel! So, common offerings are *pan de muerto* (bread of the dead) – a sweet bread decorated with bones and skulls. Warm drinks like hot chocolate are paired perfectly with sweet sugar skulls.

THE STATS

Location: Mexico City and across Mexico, particularly Oaxaca
Date: Nov. 1 and 2
Iconic features: Sugar skulls, orange marigold flowers, and ofrendas (offering tables)

DEATHLY SMELLS

The marigold flower is a common sight during the Day of the Dead Festival. This orange flower is sometimes called the *flor de muerto* (flower of the dead), as it is thought to attract souls to their family's offerings due to its strong scent and bold color.

OLD AND NEW

In 2008, UNESCO officially recognized this festival as an important cultural tradition. However, its roots go back a long time. All Soul's Day is a Spanish custom that came to Mexico when Europeans invaded in the 1500s. And the Aztec, who ruled this part of the world before the Spanish arrived, believed the dead lived on in an underworld called Mictlan. So while the festival has only recently gained international recognition, taking time to celebrate ancestors is a long-running tradition.

MORE FUN!

There's plenty more fun to enjoy in Mexico City, including its amazing modern art museum (Museo de Arte Moderno), where you can check out artworks by Diego Rivera (he of calavera fame), Frida Kahlo, and more. Farther afield are numerous remains of the many ancient civilizations that used to thrive here, including the Maya and the Toltec. Just north of the city are the great stone pyramids of Teotihuacán (right), which were built nearly 2,000 years ago.

Admire aquatic art and flying fish...

...AT GRENADA'S UNDERWATER SCULPTURE PARK

Take a trip to Grenada in the Caribbean and visit an art gallery like no other. The world's first underwater sculpture park can be found just off Molinere Bay on the island's west coast. This unique attraction is best viewed when wet, so grab your scuba or snorkel gear and jump into the warm Caribbean Sea.

PERFECT PLAYGROUND

This underwater garden was created by British sculptor Jason deCaires Taylor in 2006. There are 75 works of art in total, situated on the seabed around 16 ft. (5 m) beneath the surface. The aim of the project is both artistic and environmental, as the works have been created from materials that encourage underwater life. Over time, corals and sponges have attached themselves to the statues, creating new ecosystems for marine life to thrive. So you won't be the only one admiring the art – schools of fish will be right alongside you too!

THE VICISSITUDES

This iconic sculpture features a group of 26 Grenadian children all holding hands in a circle. Some people think it represents the circle of life, or how children adapt to new environments, as the sea embraces them in their new underwater home. What do you think?

THE STATS

Location: Molinere-Beauséjour Marine Reserve, Grenada
Best time to visit: Dec.–May for the dry season
Size: 8,600 sq. ft. (800 sq km)
Depth: 16–26 ft. (5–8 m)
Number of artworks: 75

DID YOU KNOW?

The sculptures are made out of pH-neutral cement and stainless steel. They were created on land before being transferred into the water by cranes, then bolted to the seabed to secure them.

CHRIST OF THE DEEP

This sculpture replicates an original found on the waterfront at Carenage Harbour in St. George's, the capital city of Grenada. It was given to the people of St. George's by the city of Genoa in Italy for helping 600 passengers and crew to safety when an Italian passenger ship, the *Bianca C*, sank in the port in 1961. Artist Troy Lewis's underwater replica was made in 2011 to mark the 50th anniversary of the sinking – but, of course, this statue was sunk on purpose.

STAYING DRY

If you don't want to get wet, don't worry, there is another way to view the art. Just hop on a glass-bottomed boat to glide above the sculptures. As Jason deCaires Taylor says, "The ocean is the most incredible exhibition space" and a boat ride makes this art accessible for all.

MORE FUN!

"Liming" is the Grenadian art of taking pleasure in doing nothing! So after a morning of underwater culture, why not relax on the beach and see if you can see any flying fish leaping out of the water above the sculptures. These fish are a common sight across the Caribbean and their long, wing-like fins help them glide above the water.

Enjoy treetop adventures...

...IN THE MONTEVERDE CLOUD FOREST

Welcome to Costa Rica, where the forest meets the clouds and adventures feel mysterious in the mist. The Monteverde Cloud Forest is a reserve area that was established in 1972. It's a unique habitat that creates a home for a vast variety of plants and animals. There are many ways to explore – why not try them all?

DID YOU KNOW?

Bosque nuboso is Spanish for "cloud forest."

HANGING BRIDGES

For a more gentle experience in the clouds, why not walk the hanging bridges? There are eight of them, spanning approximately 1.8 miles (3 km). Ranging from 40–200 ft. (12–60 m) above the ground, this is your chance to spot some big cats that call the forest home, like the ocelot, puma, and jaguar.

ZIP WIRES

If you'd like flying through the clouds, then zip-lining should be your starting point. You'll be harnessed up and then guides will help you explore a 2-mile-long (3.2 km-long) course over a couple of hours. Fifteen different zip wires and 18 platforms will take you as high as 500 ft. (150 m) above the ground, so you are likely to spot at least one of the 400 species of birds that live here. You can then finish the flight with a Tarzan rope swing... if you're brave enough!

SKY TRAM

After a morning of energetic exploring, you may choose to climb farther into the forest in the privacy and comfort of your own gondola. Take the sky tram to climb one of the highest points in Monteverde, and after a 20-minute ride you'll arrive at the observation deck for a breathtaking view of this special reserve.

WILDLIFE EVERYWHERE

There's so much wildlife to spot here up in the foliage. It's best to do it in the company of a trained guide who'll know exactly where to look. Animals to check off your list include two-toed sloths, tree-dwelling anteaters called tamanduas, capuchin monkeys, tree frogs, hummingbirds, and – much more elusive and harder to spot – the resplendent quetzal.

Hummingbird

Two-toed sloth

Capuchin monkey

Tree frog

Resplendent quetzal

Tamandua

FOREST FORMATION

The cloud forest exists because of its mountainous location high up above the Pacific Ocean on one side and the Caribbean Sea on the other. Strong winds create lots of cloud cover that the Sun finds it hard to break through, so less water evaporates from the trees. The moisture trapped below the clouds provides ideal conditions for plants to grow and plenty of nutrients to support a huge amount of wildlife.

THE STATS

Location: Monteverde Cloud Forest Biological Reserve, Costa Rica
Best time to visit: Dec.–Apr. for the dry season
Size: 16 sq. miles (42.25 sq km)
Height: 5,000 ft. (1,500 m) above sea level
Walking trails: Over 8 miles (13 km)

MORE FUN!

With around a quarter of its land protected from development, Costa Rica has gained a reputation as a peaceful eco paradise. As well as enjoying its jungle wildlife, you can go whale-watching off its coast and explore live volcanoes, such as Mt. Arenal (below) – from a safe distance, of course!

Next it's...

...SPECTACULAR

SOUTH AMERICA

South America has a distinct flavor. Firstly, it's full of fun – you can celebrate at the world's biggest party at the Rio Carnival and shout your heart out at an Argentine soccer match. Secondly, it's definitely unique – you can marvel at the most unusual animals on the planet in the Amazon Rainforest and on the Galápagos Islands. But South America is also mysterious – you'll get to encounter the driest place on Earth as well as a "lost city" high in the Peruvian Andes. So take a deep breath as you explore this breathtakingly beautiful landmass.

4

PLACES TO VISIT:

1. **Totumo Mud Volcano**, Colombia, p.50
2. **Enchanted Lagoon**, Isla Grande, Colombia, p.51
3. **Angel Falls**, Venezuela, p.52
4. **Galápagos Islands**, Ecuador, p.40
5. **Amazon Rainforest**, Brazil, p.44
6. **Machu Picchu**, Peru, p.46
7. **Rio de Janeiro**, Brazil, p.42
8. **Atacama Desert**, Chile, p.56
9. **Yerba Loca National Park**, Andes, Chile, p.54
10. **Santiago**, Chile, p.55
11. **Buenos Aires**, Argentina, p.48
12. **Península Valdés**, Patagonia, Argentina, p.58
13. **Los Glaciares National Park**, Argentina, p.59

GORGEOUS GEOGRAPHY

In South America, you'll find many landscapes, from mountains and grasslands to long, craggy coasts. The most famous region is the Amazon. Here, a huge river gives life to a rainforest that covers most of the north of the continent.

DID YOU KNOW?

There are 12 countries in South America. Brazil is the biggest; Suriname is the smallest.

Explore a land that time forgot...

...ON THE GALÁPAGOS ISLANDS

Take a flight from Ecuador to the volcanic Galápagos Islands. Left untouched for thousands of years, this is a place where wildlife runs the show and some of the world's most amazing animals call home. There are so many islands to explore – and unique creatures to protect – so reserve a boat tour and let the certified guides be your gurus.

ISABELA'S HIGHLIGHTS

There are 13 main islands on the Galápagos, but over a hundred smaller islets and rocks. The largest of all is Isabela Island, formed by the joining together of six volcanoes, of which all but one are still active today. In fact, much of Isabela's highlights are a result of volcanic activity. The Cabo Rosa Tunnels (above) are a must-see – they are lava platforms that collapsed into the sea, creating amazing tunnels and bridges, both above and below the water.

GIANT TORTOISES

The Galápagos giant tortoises are the islands' most famous residents. They are the most long-lived reptiles on Earth, with a lifespan of well over 100 years. There are several different species, but two main types: those with dome-shaped shells and those with saddle-shaped shells. The difference perfectly demonstrates how life on the Galápagos Islands has been left to adapt to its environment. A saddle-shaped shell allows the tortoises to stretch their necks higher, enabling them to access different plants from the dome-shelled tortoises, which feed on foliage closer to the ground.

DID YOU KNOW?

It was seeing how the animals had adapted to their environment on the Galápagos Islands that led famous naturalist Charles Darwin to write his groundbreaking book about evolution, *On the Origin of Species*, in 1859.

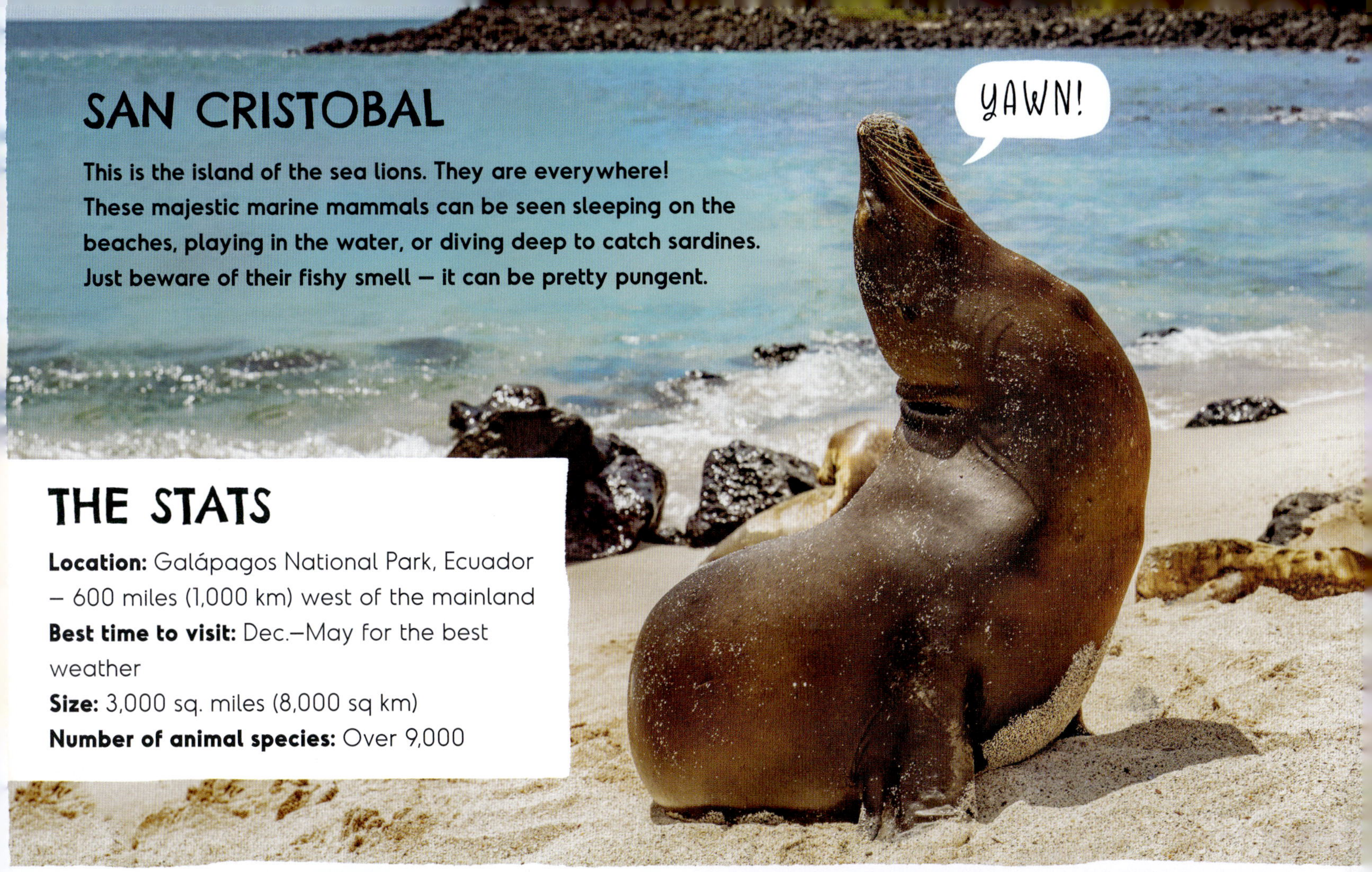

SAN CRISTOBAL

This is the island of the sea lions. They are everywhere! These majestic marine mammals can be seen sleeping on the beaches, playing in the water, or diving deep to catch sardines. Just beware of their fishy smell – it can be pretty pungent.

THE STATS

Location: Galápagos National Park, Ecuador – 600 miles (1,000 km) west of the mainland
Best time to visit: Dec.–May for the best weather
Size: 3,000 sq. miles (8,000 sq km)
Number of animal species: Over 9,000

KICKER ROCK

Take a boat trip from San Cristobal to Kicker Rock, also known locally as Leon Dormido (sleeping lion). Here you can swim between two imposing volcanic ash rocks that look like sleeping sea lions. The narrow channel between the rocks is full of fish, turtles, rays, and hammerhead sharks. The rocks are a refuge for seabirds such as blue-footed boobies (above) and the incredible marine iguanas (right). These iguanas are the only lizards to spend much of their life at sea, diving down to depths of 65 ft. (20 m) to graze on algae.

TOO MUCH ATTENTION

With fame comes frenzy, and the Galápagos Islands are a tourist hot spot these days. Increasing human presence is putting pressure on the wildlife, so conservation is a key topic for this destination.

Join the world's biggest party...

...AT THE RIO CARNIVAL

Experiencing Carnival in Rio de Janeiro, Brazil, is like turning on all your senses at once and having the best party of your life. It's held every year before the Christian period of Lent starts and originated as the chance to have one last day (or a few days!) of fun before cutting back on indulgences in the run-up to Easter. So get your glad rags on and learn how to samba as Brazil shows you how to party in style.

DANCE COMPETITION

Carnival is one big celebration of Brazilian culture, but it's also a chance for groups to show off their creative skills as they compete in a dance showcase competition. The dance is samba, an energetic ballroom dance that has lots of hip rolls, bounce, and fun. It can be danced in pairs or solo and, when people samba during Carnival time, their costume is just as important as their moves.

SAMBADROME

The Carnival dance competition is held in the Sambodrome stadium in Rio. It's one long runway with seating either side for 90,000 spectators! The performances are organized by clubs known as samba schools and are made up of numerous elements. As well as the dancing and music, each samba school creates incredibly elaborate costumes and a large float, often in the shape of an animal. Judges award the schools points, and the winner is named the champion of the Carnival.

THE STATS

Location: Rio de Janeiro, Brazil
When: Friday before Ash Wednesday until Ash Wednesday
Began: 1723
Attendance: 2 million people a day

KEY CEREMONY

When Carnival begins, the mayor of Rio hands over the keys to the city to someone playing a character called King Momo, a mythical being who leads the party. See if you can spot him when in the Sambadrome!

SENSATIONAL RHYTHMS

The country's national dance was created in Rio de Janeiro in the early 17th century. Samba is a dance that bounces along to drums and handclaps and is energetic and colorful. Although created in Brazil, it has lots of African influences. Why not give it a go? Just alternate your weight with each step to the rhythm of "quick, quick, slow, repeat." At the end, throw back your head and extend your arms out to the side!

CREATE A COSTUME

Carnival is all about dressing up! The costumes on show are designed according to a theme the samba school chooses. Every individual sequin, feather, and bead is hand stitched and so the most detailed outfits can cost up to US$10,000. Each wing of the parade has a different costume, and you can actually buy your own costume if you want to join the ground-level part of the parade. But of course, if you want to just join a street party, dress how you wish and go to town with accessories and body paint!

MORE FUN!

Outside of Carnival time, Rio is still a city packed with fun. You can sunbathe (or play soccer or volleyball) on the world-famous Copacabana beach, take a cable car to the top of Sugar Loaf Mountain, or enjoy fantastic views out of the city from Christ the Redeemer (right), one of the world's most famous statues.

Find animals everywhere...

...IN THE AMAZON RAINFOREST

It's no surprise that the largest rainforest in the world is on your ultimate travel list. Everyone knows it's jam-packed full of wildlife! But as it spans eight countries, contains around 400 billion trees, and is home to 10 percent of all species on Earth, knowing how and where to begin exploring it can be daunting. To try and see as much wildlife as possible, here's one way to discover what this wild slice of South America has to offer.

RIVER CRUISE

Running through the giant Amazon Rainforest is the mighty Amazon River. Taking a tour on a cruise boat is a great way to introduce yourself to both. You can also explore the river using a kayak or paddle board. Look out for wildlife both in the riverside trees and the water itself, which can include pink river dolphins, giant otters, and manatees. Beware of the predatory piranha, however, a meat-eating fish with fearsome teeth. You should also make a hasty exit if you spot the nostrils of an anaconda peeking above the water – as the largest snake on Earth, the green anaconda can grow up to 20 ft. (6 m) long!

TIME FOR A TREK

Once the boat moors up, it's time to put your walking boots on and join a tour. You're likely to get lost if you head out on your own! One of the iconic mammals you definitely want to try and spot is the jaguar – although it'll be difficult, as they like to stay hidden. A top rainforest predator, this big cat is powerful and strong, and can be seen in the trees as well as in the water.

DON'T TOUCH!

An Amazon experience wouldn't be complete without a bit of danger. Some of the animals that live here have adapted interesting features to help keep them safe. Poison dart frogs (below) will hopefully be easy to spot on your trek as they have amazing, brightly colored bodies. It's their way of telling everyone not to touch – they are poisonous! Eyelash vipers are another example of animals with a warning sign: yellow, red, green, or pink skin. Their venom can be fatal, but luckily they are not usually aggressive.

LOOK TO THE SKIES

Much of the life in the rainforest exists high up among the leaves and branches of the canopy. So look up to see numerous monkey species, as well as some record-breaking birds. The white bellbird is the loudest in the world, while the Spix's macaw is the world's rarest. The dinosaur-like hoatzin is found only around the streams and lakes of the Amazon River. But it is the toucan that's become the symbol of this rainforest, particularly the colorful keel-billed species (below). Its bright bill is almost as long as its body!

THE STATS

Location: Brazil, plus Bolivia, Peru, Ecuador, Colombia, Venezuela, Guyana, Suriname, and French Guiana

Best time to visit: Jul.–Dec. for the dry season

Size: 2.5 million sq. miles (6.7 sq km)

Length of Amazon River: Around 4,000 miles (6,400 km)

Number of trees: Around 400 billion

FAMOUS FLORA

Of course, the animals of the Amazon have such a fantastic place to live because of all the plants that grow here, particularly the trees. Huge 165 ft. (50 m) tall rubber, cacao, and Brazil nut trees are just some of the fabulous flora that give the rainforest its structure. But life can be found at all levels. On the dark, humid forest floor, keep your eyes peeled for some special creepy-crawlies, including leafcutter ants (left), bird-eating spiders the size of dinner plates, and big, plump Hercules beetles.

LOSING TREES

Deforestation is sadly a huge problem for the wildlife of the Amazon Rainforest. With an estimated area as big as a soccer field being chopped down every minute, we may never discover some of the incredible species that live here unless we act quickly. Deforestation is not good for our planet either. When trees are cut down, the amount of carbon dioxide in the atmosphere increases, which in turn affects global warming, as carbon dioxide is a greenhouse gas.

Discover a lost city...

...AND TREK TO MACHU PICCHU

Imagine a mysterious mountain fortress, the home of a powerful emperor of the vast Inca civilisation. This incredible "city in the sky: lay abandoned and forgotten about for several hundred years. Could you be the explorer to find it and unearth its secrets? Bring this story to life by travelling to Cuzco in Peru, then trekking along the Inca Trail to Machu Picchu, a real lost city that lay hidden from the outside world until 1911.

PICTURE PERFECT

Hiking the Inca Trail to Machu Picchu is an adventure of jaw-dropping beauty. Snowcapped Andean mountains and low-lying clouds add to the mystery of this ruined city. Positioned between two peaks and surrounded by forests, access can be tricky, but people still flock here to see the spectacular sight. It's no wonder that it was named one of the New Seven Wonders of the World in 2007.

THE STATS

Location: Andes Mountains, Peru

Best time to visit: May–Sept. for the dry season

Size: 325 sq. miles (325 sq km)

Height: 7,970 ft. (2,430 m) above sea level

Built: Mid-15th century

INCA TRAIL

Take on the classic four-day Inca Trail trek to get the best experience. After leaving Cusco – the ancient Inca capital – day one involves a relatively flat start, with a crossing of the Urubamba River. Remember to look out for Andean condors above you. On day two, you'll climb higher, through a cloud forest. Day three involves heading up to Warmiwañusqua Pass to enjoy a beautiful panoramic view. As you edge closer to the citadel of Machu Picchu itself, you'll find more ruins, an Inca altar, and stone trails to transport you back in time.

CITY SIGHTS

It's believed that Machu Picchu was built as a royal retreat – a sort of vacation home for the emperor – but it was probably also a religious site. The citadel is divided into a farming area and an urban area. Crops were grown on terraces cut into the mountainside, while houses, plazas, baths, and temples make up the urban section (below). The site puzzled archaeologists for a long time – how did the Inca people transport the stones up the mountain? It's now believed they were literally pushed up!

SUN GATE

After four days of walking, the trail comes to an end, and you'll reach a series of steep steps that lead to the magnificent Sun Gate (above). Follow in the footsteps of the Inca emperor himself as you pass through the gate, the citadel coming into view. If you're here at sunrise, you'll be treated to a magical picture-postcard view. It's thought that the gate was created to honor the Inca god of the Sun, Inti.

LLAMA LOVE

These mammals can been seen grazing on the grass all around Machu Picchu, along with alpacas. They were important to the Inca for their wool and meat, and were also used to help carry heavy loads. Today the llamas mainly like to photobomb the tourists!

Wave your flag...

...AT A BUENOS AIRES SOCCER MATCH

Enjoy watching soccer? Believe you're a supportive fan to your team? Jump out of your seat when a goal is scored? This is pretty typical soccer fan behavior, but it's time to take it to the next level and experience the energy of being a fan of an Argentine team. A match in Argentina has an atmosphere like no other, and with 24 professional teams in one city, the capital Buenos Aires is the place to be!

LA BOMBONERA

Every socce3r fan knows that La Bombonera, the stadium of Boca Juniors, has an electric atmosphere. Of course, fans help create the noise in a match, but La Bombonera has been designed to have incredible acoustics. Built in the shape of a "D," it has steep sides, and the seats run right alongside the field, so fans are really close to the players. All of this helps increase the sense of excitement before the game has even kicked-off.

SUPERCLÁSICO

When Boca Juniors play against opponents River Plate at La Bombonera, journalists describe it as one of the most intense sporting events in the world. These two clubs have been suburb, city, and national rivals since the 1900s when they were both founded in the La Boca neighborhood. River Plate moved to the north of the city in 1925 and has been nicknamed "Los Millionaires" (the millionaires). Boca Juniors is seen as the club of the working class, and the team prides itself on its grit and determination to win.

DID YOU KNOW?

A goal is not just a goal in Argentina. You've got to sound out the "O" for as long as possible to really celebrate each and every score!

World Cup winner Diego Maradona used to play for Boca Juniors.

SONG AND DANCE

Argentine supporters show their love for their team by singing and dancing. Each club has its own chants – you'll soon pick them up. Friendship and loyalty are other key characteristics of Argentine soccer supporters, and the crowd is behind their team no matter what happens. So pick a team and sing your love for them!

THE STATS

Location: Alberto J. Armando La Bombonera Stadium, La Boca, Buenos Aires, Argentina

Best time to visit: May–Dec. for the soccer season

Club founded: 1905

Stadium opened: 1940

Number of spectators: Approx. 54,000

Championships won: 44 national, 18 international honors

MORE FUN!

While in the city, you also need to check out a performance of tango dancing (above right). The art form originated in Buenos Aires in the late 19th century, and has become a vital part of Argentine culture. For a different sort of spectacle, Buenos Aires is also home to the world's widest avenue (according to Guinness World Records). Avenida 9 de Julio (right) measures an incredible 360 ft. (110 m) across and has 16 lanes of traffic.

Slip, slide, squelch, and wallow...

...INSIDE THE TOTUMO MUD VOLCANO

Welcome to Totumo Mud Volcano, a must-visit when in Cartagena, Colombia. This volcano may be the smallest active volcano in the country, but it certainly attracts the most visitors. Instead of lava bubbling in the crater of this volcano, you'll find warm mud! And as the mud is alleged to have healing properties, people can't wait to jump in and take a bath.

SQUELCH!

MUD, GLORIOUS MUD

You may wonder why people ever thought jumping into a mud-filled volcano was a good idea. It doesn't sound very appealing, after all. Volcanoes mostly occur where plates in the Earth's crust meet and magma works its way through the gaps to the surface, erupting as molten lava. Instead of lava, this volcano contains hot spring water and ash, which has mixed together to create thick, warm mud. Containing rich mineral deposits from deep underground, this is no ordinary mud – it's been supercharged by the volcano! The brown sludge is packed with natural minerals such as magnesium, sodium, and calcium, which are thought to have beneficial effects on the body's joints, muscles, and skin.

LOCAL LEGEND

The story goes that the Totumo Mud Volcano used to spew out hot lava and ash – just as any other erupting volcano would. The locals, however, believed this was the work of the devil, so asked a priest to help. He decided to banish the devil from the volcano by sprinkling holy water into the mound, turning the contents into a soothing mud bath.

SINK OR FLOAT?

Totumo Mud Volcano is 50 ft. (15 m) high and accessible via a wooden staircase. But don't worry about sinking down that distance once you're in the mud – it's so dense, it's easy to float on top. Once at the top of the volcano you will have to climb down a ladder before squelching into the thick, gray-brown sludge.

WASH IT OFF

When you're ready to leave the volcano, you just need to climb back up the ladder. But this is easier said than done as you follow in the footsteps of multiple other mud-clad humans who have left a slippery trail behind them! Luckily, help is on hand when you make it out of the volcano. Locals will help you wash off using water from the nearby lagoon. But despite how clean you feel after your dip in the sea, you are likely to find mud on you for many days to come!

ENCHANTED LAGOON

Once you're back from your trip to the mud volcano, how about an evening swim? You can go one better at the Enchanted Lagoon on the island of Isla Grande near Cartagena... you can swim in a sea of stars! This is all down to bioluminescent plankton in the water, which light up as you swim past and disturb them, looking like hundreds of fairy lights twinkling in the water. If it's an especially dark night, you're in for a magical fairy-tale show!

DID YOU KNOW?

The plankton turn a bright blue due to a chemical reaction that happens inside them. Scientists believe this happens to distract predators and attract prey.

THE STATS

Location: Near Cartagena, Colombia
Best time to visit: Dec.–Mar. for the dry season
Height: 50 ft. (15 m)
Age: Approx. 5,000 years
Capacity: Up to 15 people at a time

Fly over the top of...

...THE WORLD'S TALLEST WATERFALL

Angel Falls, the highest uninterrupted waterfall in the world, is found deep in the rainforest of Venezuela. Cascading down Auyán-Tepuí mountain, its beauty has to be seen to be believed. But, as getting to the falls across land and water can take days of travel, why not see them as their namesake first did... from directly above?! A helicopter ride will show you this natural wonder in all its glory.

BIG SCREEN SCENES

The Canaima National Park in southern Venezuela is a UNESCO World Heritage Site. Lush rainforest surrounding tabletop mountain formations makes for breathtaking scenery. In fact, it's so stunning it's inspired scenes in movies like *Up* (below right) and *What Dreams May Come*.

DID YOU KNOW?

Angel Falls is 20 times taller than the mighty Niagara Falls!

THE STATS

Location: Venezuela
Best time to visit: Aug.–Sept. when the falls are at their biggest
Height: 3,212 ft. (979 m)
Drop: 2,648 ft. (807 m)

FIRST SIGHT

As you fly toward this magnificent, otherworldly scene, the falls come into view as a dazzling thin line cutting down the center of the rock face. To ascend the 3,212 ft. (979 m) to the top of the mountain, you'll need a head for heights. The sheer drop off the cliff edge is intense, but the view out across the national park is worth it.

RIVER RIDE

If you're not enthusiastic about the kinds of heights you'd experience in a helicopter, there's another way to see Angel Falls – although it is a little tricky. With no roads in sight, you'll first need to travel by airplane to Canaima National Park and camp overnight. The next morning, you'll be up early to take a boat or canoe to reach the base of the falls. This is only possible for around half of the year as the river needs to be deep enough, so plan your trip carefully to avoid getting stranded in shallow water!

NAME GAME

Known as *Kerepakupai Merúin* in the local Indigenous language, the waterfall is more widely associated with Jimmie Angel, an American pilot who spotted it as he flew overhead in 1933. When he came back later to try and land above the falls, his plane (shown here) was damaged, and he had to spend 11 days climbing down the mountain. Angel's story was so inspiring that the falls were named in his honour, and his ashes were scattered over them in 1960.

Ride on horseback with Chilean *huasos*...

...OVER THE ANDES

Ready to become a huaso for the day? That's the name given to skilled Chilean horseback riders who roam the vast countryside herding cattle and farming the land. Famous for their bravery and skill, their culture has become legendary in folklore and literature. So join your tour group for the day, saddle up, and trot off into the wilds for this truly South American experience.

DRESS THE PART

A traditional *huaso* wears a poncho called a *manta* and a straw hat called a *chupalla*. Underneath the poncho, the rider's vest and leggings have loops and pockets to hold their tools. A huaso's saddle not only serves as a seat and storage space, but also as a blanket and a bed thanks to the layers of material folded carefully underneath it.

EXPLORE THE HIGHS AND LOWS

The Andes are the world's longest mountain range above water, stretching almost the full length of South America. Their majestic peaks, varying temperature ranges, and high plateaus offer a variety of habitats for plants and animals. To the east of the mountains are the pampas, a vast area of low-lying grasslands with rich soil, which are ideal for cattle and are home to the huasos. As you ride across the plains with your group alongside you, the backdrop of snowcapped mountains makes for a picture-perfect view, especially when the Sun sets behind them!

THE STATS

Location: Yerba Loca Natural Park, one hour drive from Santiago, Chile
Best time to visit: Sept.–Mar. for the best weather
Size: 150 sq. miles (390 sq km)
Open: 8:30 a.m.–6:00 p.m. daily
Length of ride: From 2.5 hrs. to a full day

FAMOUS IN FOLKLORE

Similar to the Argentine gaucho, the huasos' culture has become legendary in Chilean folklore and literature. These figures are admired for living independent lives, free to roam wherever they wish across a spectacular landscape. They play an important part in many parades, festivals, and holidays. Indeed, September 17 has been named "Huaso Day" in their honor.

FROM FOUR LEGS TO TWO

Riding around the Yerba Loca Natural Park, you will have spotted the region's numerous mountains. While horses can get you to most places, you may also wish to explore the area on foot. There are various hiking trails to viewpoints where you can gaze out over the Andes mountains and La Paloma Glacier. The more adventurous could take on the one- or two-day hike up to the summit of the mountain, Cerro Manchón, at 12,267 ft. (3,739 m) above sea level. In winter, if there's enough snow, people even ski-hike to the summit here in a sport called randonée skiing.

MORE FUN!

The capital of Chile, Santiago sits in a valley surrounded by the Andes. Head up to the Parque Metropolitano in San Cristóbal by cable car (left) for the best views in town. In summer, you can also cool off in two public swimming pools: Tupahue and Antilen. If you'd like some space to swim, then you should also visit one of the largest pools in the whole world at San Alfonso del Mar – it's just 62 miles (100 km) west of Santiago.

Stargaze in the driest place on Earth...

...AT THE ATACAMA DESERT

This experience may look a little dull at first, as there's not much in the way of plant or animal life in the driest place on Earth – the Atacama Desert. If you know where to look, you'll find sparkling salt flats, NASA space-testing sites, and extremely clear skies overhead, meaning stargazing opportunities that are out of this world!

THE STATS

Location: Northern Chile
Best time to visit: Dec.–Feb. for the clearest skies
Area: Approx. 50,000 sq. miles (80,000 sq km)
Length: Approx. 700 miles (1,100 km)
Height: 5,000–6,560 ft. (1,500–2,000 m)
Annual rainfall: 0.6 in. (15 mm), though some weather stations have never recorded rain

EARTH'S DRIEST PLACE

This desert is actually a plateau 620 miles (1,000 km) long, just west of the Andes mountain range in Chile. The reason it is so dry is because of its position between two mountain ranges, the Andes and the Chilean Coast Range – these two rocky barriers literally block any moisture reaching the plateau. Weather stations can be found dotted around the Atacama, and some have never recorded even a single drop of rain!

WISH UPON A STAR

With almost no settlements or artificial lights – and very few clouds – the skies here are among the best in the world for stargazing. The world's largest collection of radio telescopes, ALMA, is located in the Atacama Desert. If you don't have a telescope, don't worry – you can see so much of the night sky with just the naked eye. Take a star map with you and figure out which groups of stars, called constellations, you can spot.

DID YOU KNOW?

During the day temperatures can reach 104°F (40°C) in the Atacama Desert, but at night they dip to a chilly 41°F (5°C).

VALLEY OF THE MOON

An area of the desert called El Valle de la Luna (valley of the Moon) has soil that is similar to samples taken from the planet Mars, as well as rock formations that look like they're from another planet. So NASA often makes use of the area to test its technology, such as Martian rovers (below). Many movies and television programs set in space are filmed here too!

SALT FLATS

The Atacama Desert is a vast area with many unusual sights. To the eastern edge, you'll find the Nevado Tres Cruces National Park. Within this is Salar de Maricunga (below), a salt flat created by the evaporation of water leaving behind minerals on the surface. Shimmering in the sunshine, it sits beneath the world's highest volcano, the dormant Nevado Ojos del Salado.

FIND A FLAMINGO

Certainly one thing you wouldn't expect to see in the Atacama Desert is a flamingo! But you may well spot an unexpected flash of pink, revealing the presence of Andean, Chilean, and James's flamingos just south of the salt flats. These birds like to hang out in shallow wetlands and at high altitudes, hence their presence in the Atacama Desert after a rare rainfall.

Meet marine wildlife from a kayak...

...IN PATAGONIA

Welcome to Patagonia, a region in the south of South America. The rivers and oceans around Patagonia are rich with marine life, and the UNESCO World Heritage Site of the Península Valdés in Argentina is the number-one destination for viewing this type of wildlife. It is home to an important population of South American sea lions, as well as elephant seals, penguins, and two species of whales. Time to explore!

LION'S ROAR

To witness some impressive sea lion hunting displays, be sure to visit the Península Valdés between February and March. With their namesake loud roar, you'll hear the sea lions before you see them! These mammals feed on a variety of fish and squid, and large males may also try their luck hunting young elephant seals or young fur seals. While the adults are out hunting, the sea lion pups gather together in a nursery and play. Very cute! Take some binoculars so you can watch them from the boardwalk.

CLOSE CONTACT KAYAKING

To get up close and personal with the sea lions, you'll need to be on the water. So a three-day kayaking adventure is a great way to do it. You'll get to explore bays and inlets along the coastline, and, as sea lions are sociable creatures, they'll soon come swimming over, bobbing up right beside your kayak. For an even closer encounter, just jump in for a quick snorkel with them! At night, set up camp on the beach and you'll get to experience the magic of the sunset, clear starry skies, and then sunrise, just as the sea lions do.

PENGUIN PARADE

At Punta Norte on the tip of the peninsula, you'll find hundreds of thousands of breeding pairs of Magellanic penguins. They are happy for you and your guide to walk among them, and if you visit in September or October, you'll likely find them laying their eggs in burrows or bushes. Then, from January or February, you'll find loads of cute chicks waddling about. Look for the double black stripe under their chin – it's their identifying feature!

ELEPHANT SEALS

These big beasts are excellent swimmers. They can dive to 4,920 ft. (1,500 m) and stay down for as long as two hours. On land, you'll often see males fighting each other for their territory and the attention of females, plus plenty of pups trying to keep out of the way. Do also keep a lookout for southern right whales that often swim very close to the shore.

ORCA GYMNASTICS

The waters off the Península Valdés offer the chance to see a unique sight: orcas (killer whales) intentionally leaping out of the water to hunt sea lion pups right on the sand. This is the only place in the world that they beach like this, usually only in March and April.

THE STATS

Location: Península Valdés, Patagonia, Argentina
Size: 1,400 sq. miles (3,600 sq km)
Best time to visit: Feb.–Apr. for wildlife watching
Best time to see sea lions: Feb./Mar.–Apr.
Best time to see orcas: Mar.–Apr.

MORE FUN!

A trip to Patagonia wouldn't be complete without seeing a glacier. The region is famous for them and the Los Glaciares National Park is where you'll find the 19-mile-long (30 km) Perito Moreno Glacier (left). From specially constructed boardwalks, you can view the mighty river of ice – not to mention hear it creak and groan as it moves.

Next it's...

...EXCEPTIONAL EUROPE

Many of the experiences available in Europe come with a historical twist. Ancient cities, like London and Paris, are here for you to explore, as are age-old castles and glittering collections of royal jewels. And with history comes mystery – will you dare to learn about vampire legends in Romania, or tour creepy catacombs in France? Happily, there's lots of modern-day fun to be had as well. Enjoy exploring toy stores, amusement parks, and even tomato-throwing festivals, before cooling off with a visit to a village in Finland where it's always Christmas, and a stay in an ice hotel in Sweden.

PLACES TO VISIT:

1. **Tromsø**, Norway, p.62
2. **Ice Hotel**, Jukkasjärvi, Sweden, p.66
3. **Santa Claus Village**, Rovaniemi, Finland, p.64
4. **Jacobite Steam Train**, Glenfinnan Viaduct, Scotland, UK, p.75
5. **LEGOLAND® Billund Resort**, Denmark, p.69
6. **Dyrehavsbakken**, Denmark, p.68
7. **Tivoli Gardens**, Copenhagen, Denmark, p.68
8. **Universe**, Denmark, p.69
9. **Alnwick Castle**, England, UK, p.75
10. **Durham Cathedral**, England, UK, p.75
11. **Warner Bros. Studio Tour**, Watford, England, UK, p.74
12. **London**, England, UK, pp.70–73
13. **Paris**, France, p.76
14. **Neuschwanstein Castle**, Bavaria, Germany, p.84
15. **Bran Castle**, Transylvania, Romania, p.78
16. **Buñol**, Spain, p.80
17. **Mount Stromboli**, Sicily, Italy, p.82
18. **Mount Etna**, Sicily, Italy, p.82

DID YOU KNOW?

Measuring just 0.2 sq. miles (0.5 sq km), Vatican City, located within Italy, is the world's smallest country.

CRAMMED WITH CULTURE

Europe may be the second-smallest continent but it's packed with history – and people. Over 200 languages are spoken here. There are lots of islands to explore, many of which are found in the Mediterranean Sea, a body of water almost completely surrounded by land.

1
2
3
4
5
6
7
8
9
10
11
12
13
14
15
16
17
18

Jump on a husky sled...

...TO CHASE THE NORTHERN LIGHTS

The Sun is setting and the temperature is dropping, but it's time to wrap up and get out into the snow for this experience. You're in Tromsø, in the far north of Norway, on the hunt for the mysterious aurora borealis, or Northern Lights. One of the most spectacular sights you'll ever see, these lights appear as ribbons of color dancing across the night sky. For the best chance of seeing them, you should travel to a secluded spot, away from artificial lights, so get a team of husky dogs to take you out into the wilderness after sunset.

DID YOU KNOW?

The lowest part of an aurora is about 80 miles (130 km) above Earth's surface. The top reaches up several thousand miles.

MAGICAL LIGHTS

The lights themselves are caused by the Sun sending out clouds of electrically charged particles, some of which are attracted toward the magnetic poles of planet Earth. As they enter Earth's atmosphere, they collide with molecules of gas, causing them to heat up and glow. The force of Earth's magnetic field pulls the lights into shimmering shapes. The patterns can be mesmerizing to watch.

HARDY HUSKIES

In the far north of Scandinavia, husky dogs pulling a sled is a common sight. It's been a popular mode of transportation for hundreds of years. These hardy dogs are ideal candidates for the job as they have speed, power, stamina, the ability to work as a team, and an all-important resistance to the cold. The huskies can pull the weight of a human at speeds of up to 12 mph (20 km/h) for hours on end. They love the physical exercise and are social dogs, so you really will feel like part of their dog team on your ride.

CHANGING COLORS

The colors of the lights depend on which molecules the solar particles collide with. Oxygen at lower levels in the atmosphere causes green, while nitrogen creates purple, blue, and pink tones. If you see red, that's oxygen at the very highest level of the atmosphere. With a snowy landscape muffling any sounds, watching the colors peacefully dance across the sky is a calming experience like no other.

THE STATS

Location: Tromsø, Norway
Best time to visit: Oct.–Mar.
Colors: Green, red, purple
Speed: Sun's particles hit the Earth at 45 million mph (72 million km/h)

MORE FUN!

Within the Arctic Circle, Tromsø is one of the world's northernmost cities – so be sure to wrap up warm. At Tromsø's Polar Museum, you can find out more about Arctic life. You can also catch boats from here to go whale-watching (above) or to head farther south to explore Norway's famous fjords (below) – steep coastal inlets formed by glaciers during the last ice age. Norway also hosts two of the most famous sled dog races in the world. Perhaps this experience will give you a taste for racing!

STEER YOUR SLED

Hop into the driver's seat of your sled and you take on the title of "musher." The musher is responsible for steering the sled and instructing the dog team. You'll need to learn the vocab the dogs understand… "Mush!" or more commonly "Hike!" means "Go," "Haw" is a left turn, and "Gee" is a right turn.

Celebrate Christmas early...

...AND MEET SANTA IN LAPLAND

What more could you ask for during the month of December than to step into a traditional Christmas card scene and have the chance to meet Father Christmas, a.k.a. Santa, himself? Travel to Santa Claus Village, just 5 miles (8 km) from Rovaniemi in Lapland, near the Arctic Circle, to receive the warmest of welcomes at this magical, snow-filled Christmas wonderland.

MEET THE LOCALS

Lapland is a huge area in the north of Finland, but it also extends into Sweden, Norway, and even Russia. It's within the Arctic Circle and is full of unspoiled nature. It may seem like few humans live here, but the Indigenous Sami people have called this area home since ancient times. They continue to keep the traditions of their people and their language alive, and live a life centered around reindeer herding.

DID YOU KNOW?

Father Christmas goes by different names around the world. He's Sinterklaas in the Netherlands, Père Noël in France, Papai Noel in Brazil, and Joulupukki in Finland.

AN EXTREMELY BUSY POST OFFICE

Santa Claus Village receives over half a million letters for Santa each year, which are sorted by the elves at the post office. But as well as receiving mail, Santa also sends out letters to children to arrive before Christmas. Why not write to him and ask for a letter to be sent to a friend?

THE STATS

Location: Santa Claus Village, Rovaniemi, Finland
Best time to visit: Dec. (of course), but you can meet Santa every day of the year
Opened: 1985, although it's been Santa's home for much longer!
Other attractions: Elf's Hat Academy, Snowman World, Mrs. Santa Claus Christmas Cottage

MAGICAL MEETING

When you first arrive in the village, head to the central square to find Santa Claus Office. This is where Santa bases himself when not delivering presents! Here you'll meet his elves – just watch out for the cheeky ones who may want to play snow games with you! Before long, you'll get to meet Santa himself. Don't forget to ask an elf to take a photo of this magical moment for you.

SNOWMOBILE SAFARI

If you'd like to see more of Lapland, then the best way to travel across the snow is on a snowmobile (below). These vehicles have caterpillar tracks over their wheels for stability, and skis on the front to help steer. You'll need a driver's license to actually drive one, but passengers can sit on a sled and get towed along behind.

REINDEER RIDES

Your experience in Rovaniemi and Lapland wouldn't be complete without reindeer. The number of these majestic beasts roaming the land is roughly the same as the number of people living here! Reindeer are herbivores – they graze on more than 350 different types of plants, and are always on the move searching for more. Take a reindeer sleigh ride around Santa Claus Village, or head out across the frozen landscape to experience what life is like for the local reindeer herders who call this area home.

Sleep in the world's coldest bedroom...

...AT SWEDEN'S ICE HOTEL

This is a bucket list travel experience that is built from scratch each year. When it's complete, you'll have your very own ice room for the ultimate winter sleepover. It may sound a little crazy to sleep on a bed of ice, in temperatures of 23°F (-5°C), but being snuggled up in a silent room with naturally calming snow and ice surrounding you means you're likely to have one of the best sleeps of your life!

ICE HOTEL

The original Ice Hotel was founded in 1989, in the Swedish village of Jukkasjärvi, 125 miles (200 km) north of the Arctic Circle. At the start of each winter, artists gather together to build a new and unique version of the hotel, leaving it to simply melt away in spring. Each room is an original piece of art crafted entirely from snow and ice, with many spaces featuring beautiful ice carvings, sculptures, and colorful illuminations.

ICE CREATIONS

Building an entire hotel requires ice, and a lot of it. Luckily the nearby Torne River freezes in winter and is perfect for the job, as the ice that forms here freezes slowly and naturally, meaning it stays nice and clear. Once the new hotel's design has been finalized, individual blocks of ice are frozen together to create the structure. It then takes a couple of weeks of solid sawing, chiseling, and filling to create the rooms – including carving those all-important beds.

PLANNING PERMISSION

Any adults wanting to design their own room need to think ahead. Plans need to be submitted to the creative board of the Ice Hotel by spring. Out of around 120–150 proposals received each year, only 15–20 are accepted. The lucky winners then need to wait for the season to change – and the temperature to plummet – before making the trip north to begin work.

ADDED EXTRAS

The finishing touches in the rooms are the lights and the bedding. Lighting brings the transparent elements of the ice designs to life. And, of course, bedding is essential for keeping warm and comfortable! Each ice bed is made up with a mattress, sheets, pillows, and reindeer skins, plus you're given a special sleeping bag for the night. But it's also important to wear thermal layers underneath your pajamas. Woollen socks and a hat help keep the warmth in too!

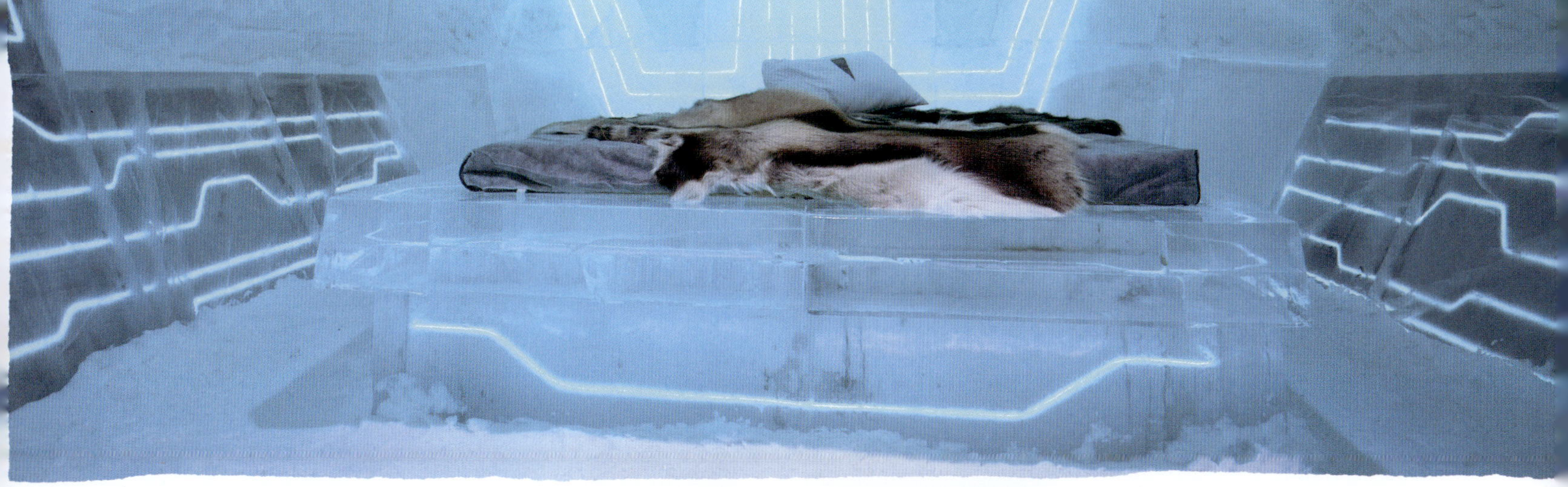

THE STATS

Location: Jukkasjärvi, Sweden
Open: 365 days – ice rooms are available from Dec. until they melt in spring/summer
Founded: 1989
Number of rooms: 53 in winter
Temperature inside: 19.4 to 23°F (-5 to -7°C)

VERY SNICE

The strongest building blocks used to create the hotel rooms are called "snice"... because they are made from snow and ice! These are very sturdy, so are best used for the floors and walls. It's been said that the amount of snow used to build the Ice Hotel equates to around 700 million snowballs!

MORE FUN!

There's plenty to do outside of your hotel too. The Northern Lights can be seen here, dancing across the sky. Husky sled rides out into the wilderness are also available, and anyone aged eight and over can take a class in ice-sculpting, learning those all-important skills to be able to make your own ice-hotel room.

Experience theme parks old and new...

...IN DELIGHTFUL DENMARK

Amusement parks have been providing fun for all the family for centuries, and Denmark has some of the original and best. Dyrehavsbakken, just 10 minutes north of the capital, Copenhagen, was founded in 1583 and is the oldest theme park in the world. You can also visit some more recent arrivals, including Tivoli Gardens (1853) and LEGOLAND® Billund Resort (1968).

TIVOLI GARDENS

Tivoli Gardens in the center of Copenhagen has been open since 1843, making it the third-oldest amusement park in the world. When visiting, you'll want to try its famous Rutschebanen wooden roller coaster, which first started operating in 1914. The Daemon ride will blow your socks off with two loops and a zero gravity roll while flying along at 48 mph (77 km/h). Tivoli is also a venue for the performing arts, so be sure to catch a concert or a pantomime. And when daylight fades, the magic increases as circus lights and lanterns light up the gardens (below).

RETRO BAKKEN

If you go down to the woods today you're sure of a big surprise... in the form of rides, entertainment, and live music! The historical Dyrehavsbakken amusement park can be found hidden in the beautiful woods of Dyrehaven, among herds of roaming deer. Like any good theme park, you're welcomed with open arms, and here they belong to the friendly, smiling figure of Pjerrot, the white-faced clown (below right). Head to the wooden roller coaster dating back to 1932 for your first thrill, then choose from over 100 other traditional attractions, such as bumper cars, circus shows, arcade games, and prize stalls.

LET'S PLAY

It's no surprise that the playful Danes are famous for one of the world's most popular toys, LEGO®, celebrated at the LEGOLAND® Billund Resort. The park was built next to the LEGO® factory in the late 1960s to both promote the toy company and to enable people to see completed LEGO® models. Nowadays, the theme park is just as much about the rides as it is the models. Roller coasters, river rapids, and a log flume sit alongside the famous Miniland (below) – a brick city based on real buildings from around the world.

DID YOU KNOW?

The name LEGO® actually comes from the Danish words *leg* and *godt* which mean "play well."

THE STATS

Location: Denmark
Best time to visit: May–Jun.
Oldest theme park: Dyrehavsbakken, founded in 1583
Oldest roller coaster: Rutschebanen at Tivoli Gardens, 1914

OUR MODERN UNIVERSE

Universe is a science park housed in futuristic-looking buildings on the island of Als (below and right). The interior is equally forward-thinking with exhibits arranged according to the park's motto: "Where fun is a science." Visit the bubble lab and energy lab, watch geysers blow, and check out the hurricane simulator and virtual-reality experiences.

Discover dinosaurs, rockets, and high fashion...

...IN LONDON'S MUSEUM QUARTER

Head on over to South Kensington in West London for plenty of adventures in the museum quarter. Come face-to-face with a dinosaur one moment, take a simulated ride into space the next, then later check out your favorite piece of fashion. With three world-class museums all within a stone's throw of each other, it's easy to get lost in these amazing collections for weeks. But as each museum is free to enter, you can pop in and out as much as you like!

THE SCIENCE MUSEUM

The Science Museum is a treasure trove of old and new. Not only will you find historically significant scientific inventions like engines, rocket launchers, telescopes, and clocks on display, but you'll also get to immerse yourself in modern science, with the chance to conduct some hands-on experiments. Make slime grow, see lightning strike, and play some of the best video games from the past 50 years. While you're here, try and track down Helen Sharman's space suit from 1991 – she was the first Briton in space.

NATURAL HISTORY MUSEUM

An iconic institution, the Natural History Museum is home to over 80 million items. The dinosaurs are its most famous residents, with skeletons of some of the largest beasts to ever roam the planet on display. Look out for Sophie – the most complete fossil of a *Stegosaurus* ever found. There are also rare gems and meteorites, marine fossils, and objects melted by lava on display.

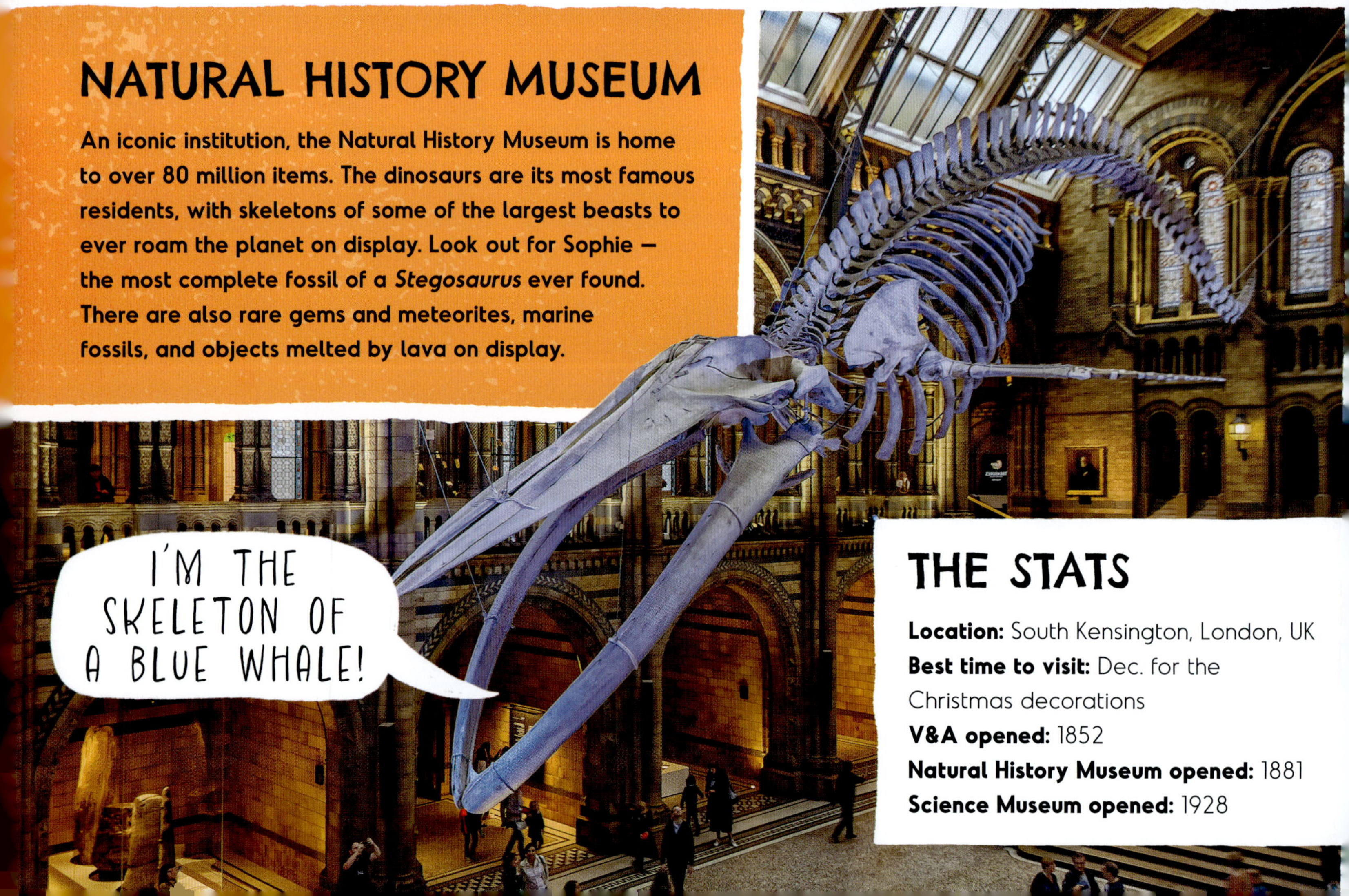

THE STATS

Location: South Kensington, London, UK
Best time to visit: Dec. for the Christmas decorations
V&A opened: 1852
Natural History Museum opened: 1881
Science Museum opened: 1928

THE V&A

The Victoria and Albert (or V&A) Museum was founded in 1852 with the mission to educate the public about art and design. It followed on from the Great Exhibition the year before, in which Londoners were treated to a showcase of everything progressive, from steam engines to exotic goods. Queen Victoria and Prince Albert presided over it all, hence their recognition in the museum's name. Today, you should look out for sculptural masterpieces from artists like Rodin, fabulous fashion and furniture, and unique oddities like "Tippoo's Tiger" – an 18th-century Indian model of a tiger eating a person. Inside the tiger is a mechanical organ.

ROYAL PARKS

After all that time spent indoors, it'd be good to get a breath of fresh air. Kensington Gardens, just down the road from the museums, is one of London's eight royal parks. While here, catch a glimpse of Kensington Palace (right), the official London residence of the Prince and Princess of Wales, then look for the statue of Peter Pan – a bronze sculpture of the "boy who would never grow up" – with fairies, mice, and squirrels at his feet, Next, visit neighboring Hyde Park, where you can hire a boat and go rowing on the Serpentine Lake.

MORE FUN!

While in London it would be rude not to visit Hamleys, on Regent Street, often called "The Finest Toy Shop in the World." Founded in 1760 by William Hamley, it is also the oldest toy shop on Earth! Staff showcase the latest toy trends across seven huge floors, so even if you don't actually buy anything, it's an experience you'll never forget. Visit during the festive season and you'll also get to view the best lights in town. People travel from all over the country to see Regent Street's impressive, dazzling Christmas lights display!

Peep at priceless royal jewels...

...IN THE

TOWER OF LONDON

Your eyes will sparkle with both delight and disbelief when you witness the whopping 23,578 gemstones kept in the Tower of London. Many of the items in this Royal Collection are still used by the monarchy for important ceremonial events today. The Crown Jewels, as the collection is known, are kept under armed guard, but all you need to do to get past is buy a ticket! How many of these highlights can you spot?

THE CULLINAN DIAMOND

Discovered in South Africa in 1905, this world-famous gemstone was the largest uncut diamond ever to be found. At 3,106 carats, it was gifted to King Edward VII on his 66th birthday in 1907. It has since been split into nine major stones and 96 smaller ones, and the largest two have been placed in the Imperial State Crown and the Sovereign's Sceptre with Cross (right), which was originally created for the coronation of Charles II in 1661.

THREE CROWNS

Keep an eye out for three extra-special crowns from the Royal Collection as you explore the historic tower. The Imperial State Crown (below left) was worn by King Charles III during the coronation procession in May 2023, and can be seen in official royal portraits. But it was St. Edward's Crown that was used at the actual moment of the crowning of the king (below right). The third crown is Queen Mary's Crown, modified especially for Queen Camilla ahead of the coronation.

ROYAL RUBY?

On the front of the Imperial State Crown you'll see the Black Prince's Ruby – which is actually a huge 170-carat red spinel. It is said to have originally been stolen by King Pedro the Cruel from the body of the Sultan of Grenada in 1362. Not long after, King Pedro was under threat, and King Edward III of England sent his son, Edward of Woodstock (the Black Prince), to help. Pedro offered treasures – including this huge jewel – to the Black Prince in return for his support. In the 15th century, the stone was worn on Henry V's helmet as he battled the French. It may have saved his life, as both the helmet and Henry V himself survived an axe blow to the head!

SOLID FORTRESS

The Tower of London was built over a period of 20 years by William the Conqueror, after his famous victory at the Battle of Hastings in 1066. It was intended to be the most secure castle in the land. The royal jewels have been stored and displayed here since 1661. They have survived many ordeals over the years, including the Great Fire of London in 1666 and an attempt to steal them in 1671.

THE STATS

Location: London, UK
Best time to visit: Jul.–Sept. when the State Rooms in Buckingham Palace are open
Visitors: Over 3 million per year
Jewelry collection: Over 100 objects

DID YOU KNOW?

The Nizam of Hyderabad necklace is valued at around US$130 million. It is probably the royal family's most expensive piece of jewelry.

RESIDENT RAVENS

Over the years, the Tower of London has been a fortress, a royal palace, a prison, an armory, a mint (a place where coins are made), a burial place and a zoo. Today, your ticket not only includes tours of the tower, battlements, and armory, but also a chance to meet the famous ravens. Legend says that the Tower of London – and the Kingdom – will fall if the ravens ever leave. Luckily, they have a dedicated Ravenmaster to care for them. He's one of the Yeoman Warders, otherwise known as Beefeaters, whose job it is to guard the tower.

MORE FUN!

If seeing all these jewels has sparked your interest in royalty, there's another place you should definitely visit – Buckingham Palace. This is the official London home of the monarch and the focal point of many royal celebrations. From July to September each year, visitors can enter the State Rooms, containing many treasures. Look for Queen Mary's Dolls' House – it's a perfect replica of a grand Edwardian house and even contains real jewels!

Go behind the scenes and explore...

...WIZARD-WORTHY MOVIE SETS

Unless you're a muggle who's been hiding under a blanket your entire life, you'll know that J. K. Rowling's Wizarding World started life as seven books, and has since grown into eight films and an amazing studio of treasures. For Harry Potter fans everywhere, visiting the Warner Bros. Studio Tour London "The Making of Harry Potter" is a dream come true.. But there are so many more places in the UK to excite true Potter fans... How many film locations can you visit?

KINGS CROSS STATION

At the start of each year, students board the Hogwarts Express to get to the Hogwarts School of Witchcraft and Wizardry. The train leaves from platform 9¾, and fans will know that Harry and his friends dash through a brick wall between platforms 9 and 10 to find it! At the real King's Cross station you can actually visit the elusive platform. Take a picture next to the disappearing luggage cart and cross your fingers that you'll get an invitation to Hogwarts one day.

THE MAKING OF HARRY POTTER

Visit the Warner Bros. Studio Tour in London and step onto sets such as the enormous Great Hall at Hogwarts and marvel at the huge feast laid out before your eyes. You'll also walk among the spooky trees of the Forbidden Forest and get the chance to do a bit of window shopping in Diagon Alley (right). You can see the original props, costumes, masks, and scale models on display, plus technical drawings and designs created by the art department. Of course, it wouldn't be magic without some special effects – check out the green screen technology that enables you to fly a broomstick!

THE STATS

Location: Warner Bros. Studio Tour, Watford, UK
Best time to visit: All year round
Total books sold: Over 600 million
Books in series: 7 since 1997
Movies: 8 (2 movies for the final book)

JACOBITE STEAM TRAIN

Riding the Hogwarts Express can become a reality if you take the Jacobite Steam Train on the West Highland Line from Fort William to Mallaig in Scotland. You'll chug over the Glenfinnan Viaduct with its 21 arches, which was made famous in the movies.

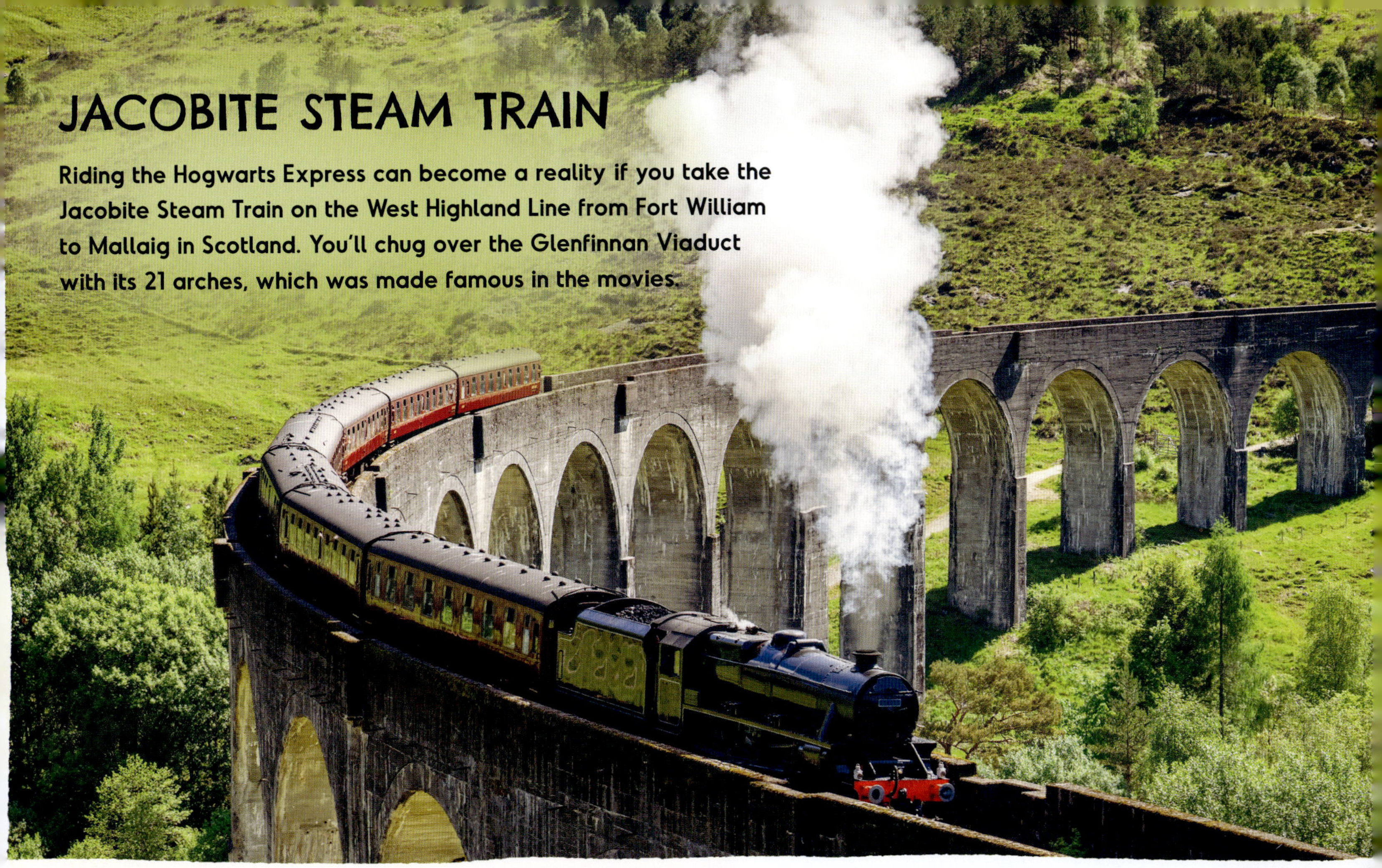

DURHAM CATHEDRAL

Many scenes showing the inside of Hogwarts were filmed in cathedrals, castles, and ancient universities. Durham Cathedral (right) has a secluded passageway called a cloister that was used as the set when Harry, Ron, and Hermione walked between lessons. A bunch of keys that Hogwarts' caretaker, Filch, carries in one movie was an actual set from the cathedral. The distinctive round Bodleian Library in Oxford is the Hogwarts' school library, and many of the cloisters and doorways of Gloucester Cathedral were used in the filming, too.

BROOMSTICK TRAINING

Feel like a flight? Harry managed to master his very first flying lesson while filming at Alnwick Castle in Northumberland. Here you can get to grips with a broomstick, master mounting and dismounting, then have a photo taken to prove you can fly! You'll zip across the lawns like you're playing a quick game of Quidditch in no time. While you're here you might think the Lion Arch entrance looks familiar, too. That's because it is – it was used as the way in and out of Hogwarts!

Keep your wits about you...

...IN THE CREEPY

CATACOMBS OF PARIS

Come on down – down into a city below a city, where 200 miles (320 km) of maze-like tunnels have existed since Roman times. The limestone removed from the ground to create these tunnels was used to build the city of Paris above them. A small section of this underground world is open to the public – the Denfert-Rochereau Ossuary, or as it's more commonly known, the Catacombs. These caves are not your average tourist attraction though – the Catacombs are actually a huge underground grave!

MUSEUM OF BONES

The bones had only been loosely piled in their new underground home when they were moved from the cemeteries, and so the Catacombs needed to be tidied before people could visit them. The French politician and scientist Héricart de Thury arranged for the bones to be built into patterned walls, where rows of leg bones were alternated with skulls. Behind these, he put the remaining bones, many of which were just small fragments. Decorative columns were also created in the tunnel network, and two cabinets displayed unusual bone specimens. It's thanks to de Thury's organization of everything that the whole area is still accessible today.

UNDERGROUND GRAVEYARD

The Catacombs house the skeletal remains of over 6 million deceased Parisians. By the late 18th century, the cemeteries in Paris were becoming overcrowded, creating a serious health risk to the living. In 1785, the authorities began removing bones from huge communal graves in the dead of night, relocating them to the tunnels. The Catacombs, as they were named in 1786, became a source of spooky interest to the locals and they were opened to the public, initially by appointment only, in 1809.

THE STATS

Location: Paris, France
Best time to visit: Apr.–Jun.
Depth of the Catacombs: 65.6 ft. (20 m)
Contains: Bones of over 6 million people
Number of visitors: 550,000 annually

DID YOU KNOW?

There are around 200 miles (320 km) of tunnels beneath Paris. The Catacombs only make up a small part of them.

GUIDED TOUR

To get into the Catacombs, you'll be guided down 131 spiral steps, taking you 65 ft. (20 m) underground. Prepare to find it a bit chilly – the temperature is around 57°F (14°C) all year round – and you'll need to watch your step as the floor can be slippery due to the humidity. It takes about 45 minutes to walk the one-way 1-mile (1.5-km) circuit, and once you're ready to exit, you'll need to climb 112 steps to get out into the open air again.

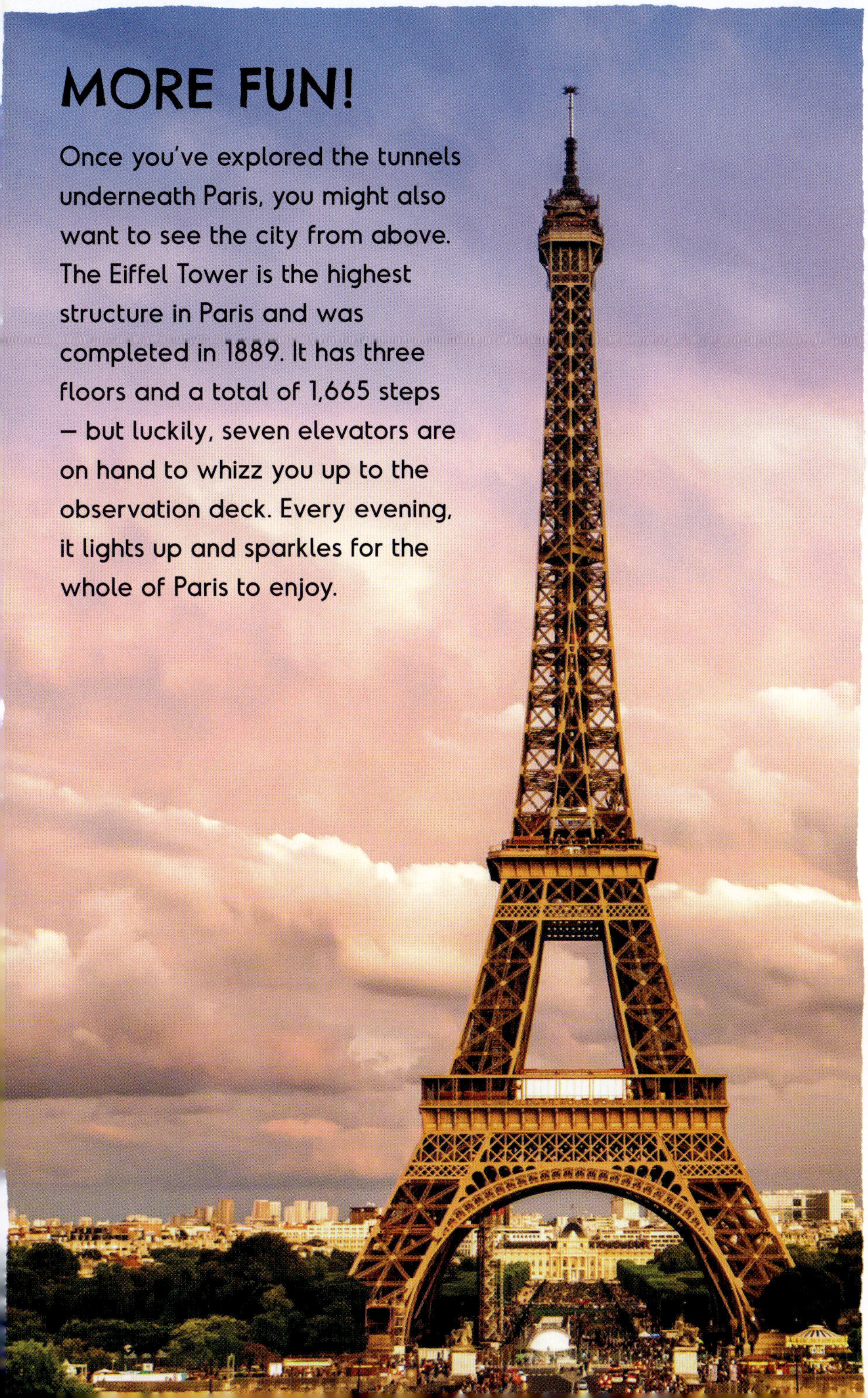

MORE FUN!

Once you've explored the tunnels underneath Paris, you might also want to see the city from above. The Eiffel Tower is the highest structure in Paris and was completed in 1889. It has three floors and a total of 1,665 steps – but luckily, seven elevators are on hand to whizz you up to the observation deck. Every evening, it lights up and sparkles for the whole of Paris to enjoy.

MOST VISITED

While in Paris, add the most visited museum in the world, the Louvre, to your travel itinerary. Many iconic pieces of art are on show here – one of the most popular being the *Mona Lisa* by Leonardo da Vinci (above). This painting is so valuable it has a screen of bulletproof glass in front of it. Which is just as well, as rocks, paint, and even soup have been thrown at it over the years by protesters wanting to make a point.

Discover if vampires exist...

...IN TRANSYLVANIA

Take a trip to Transylvania to find out if vampire legends are fact or fiction. Located in Romania, Transylvania is the place most often associated with tales of vampires. Count Dracula is the most famous fictional vampire around, but perhaps there's some reality to it all as Bran Castle in the region has plenty in common with the count's residence! There's lots of intriguing information to uncover, so you're certainly in the right place to investigate these mysterious creatures, and decide for yourself whether vampires are real, or not...

WHAT IS A VAMPIRE?

A vampire is a human-like creature that preys upon people and sucks their blood with its long fangs – eek! Vampires are thought of as being deceased humans that rise from their coffins and come to life again when night falls. While some legends talk of their horrible appearance, others tell of their amazing beauty. You'd have to see one for yourself to know for sure, but these beasts have never been caught on camera – could you be the first to capture a photo of one?

LOCAL LEGENDS

Fictional characters like vampires often begin life as local legends. In many Romanian villages, people used to believe in *strigoi* – troubled spirits who had risen from the grave. They could transform into animals or become invisible, and they gained energy from drinking the blood of their victims.

DID YOU KNOW?

Vampires hunt at night, but when the first light appears again, their power to harm people disappears.

CASTLE LIFE

Bran Castle is located between Wallachia and Transylvania, which was in Hungary when building work started on the castle around 1377. Over the years, it was occupied by various Hungarian kings and noblemen before Transylvania became part of Greater Romania in 1918. The castle was then inhabited by the Romanian royal family until they were forced to leave the country in 1948. By 1993, Bran Castle was fully restored, and has been open to the public ever since. When you visit, it may be best to take some garlic with you, just in case – it's believed to weaken a vampire's spirit!

NOVEL IDEA

Bram Stoker's novel *Dracula* was first published in 1897. Stoker's description of Dracula's castle is very similar to that of Bran Castle – a Gothic castle, perched high on a rock. Stoker's castle is home to Count Dracula, a charismatic nobleman who wishes to relocate to England. Once there, Dracula spreads a curse by biting people and turning them into vampires too. It isn't long before people start to get suspicious and set about hunting this strange figure down.

THE STATS

Location: Bran Castle, Wallachia near Transylvania, Romania
Best time to visit: Jun.–Aug. for the best weather, or Jan., when it's at its darkest and most mysterious
Height: 2,500 ft. (760 m) above sea level
Built: Late 14th century
Number of rooms: 57

NAMESAKE

In the old Romanian language, the word *dracul* means "dragon" and *dracula* means "son of dragon." Bram Stoker based the character of Dracula, who likes to impale his victims with his fangs, on a vicious ruler from Transylvania who liked to impale his victims on wooden spikes: Vlad III – or "Vlad the Impaler." Vlad III's father, Vlad II, was called Vlad Dracul, which meant Vlad III also had the name Vlad Dracula – perfect inspiration for Stoker's evil fictional character!

MORE FUN!

Other attractions to check out while in Romania include the capital, Bucharest, home to the world's heaviest building, the 3,100-room Palace of the Parliament. In the east of the country is one of the continent's best areas for spotting wildlife, the Danube Delta, marking the place where the mighty river pours into the Black Sea. It's home to hundreds of species of birds, including eagles, herons, and pelicans (right), as well as mammals such as wild boar and otters.

Prepare to get splatted...

...AT

LA TOMATINA FESTIVAL

Now for a fun experience you're never going to forget – a festival like no other. Put simply, this must-do Spanish event is one big food fight! The food is tomatoes, and the fight is 20,000 people throwing 160 tons (145,000 kg) of tomatoes at each other in the town square of Buñol, near Valencia, Spain!

ON YOUR MARKS

Nowadays the festival is so popular, the partying lasts a week. With a paella contest, fireworks, music, parades, and tomato-throwing events for both children and adults, there are many ways to enjoy the celebrations. On the morning of the main event, before the first tomato is thrown, the Palo Jábon event takes place. A large piece of ham is placed at the top of a long pole, and the pole is covered in grease. The aim is to climb up and topple the ham, allowing the tomato throwing to begin. Climbing a slippery pole is not as easy as it sounds, so a water cannon is also on standby to signal the beginning of the festivities in case no one manages to climb it!

UNUSUAL ORIGINS

This unique festival came about quite by accident! In August 1945, at a parade in the town square of Buñol, it is said that part of a performer's costume got damaged. In a rage, the performer reached for tomatoes on display at a nearby market stall and started a food fight. The following year, locals decided to reenact the fight for fun with their own tomatoes. Although the police quickly broke the fight up, it heralded the start of a new tradition. Despite the authorities trying to ban the event over the next few years due to its lack of religious importance, the townspeople protested, and La Tomatina festival became an official celebration, occurring every year on the last Wednesday in August.

THE STATS

Location: Buñol, Spain
Best time to visit: Last Wed. in Aug. for the main festival; last Sat. in Aug. for the kids' festival
Visitors: Around 20,000
Duration: Main festival – 1 hour; kids' festival – 30–40 minutes
Age range: La Tomatina Kids is for 4–14-year-olds

LA TOMATINA KIDS

Created in 2013, La Tomatina Kids offers the same tomatoey fun as the main event, but is only for children. If you're aged between 4 and 14, grab some old clothes and a towel and head to Buñol's town square on the last Saturday in August for a very messy battle and a lot of laughs!

RULES OF PLAY

Throwing squishy, overripe tomatoes may not seem like it needs rules or regulations, but when thousands of people are involved, it definitely does. It's important to squash the tomatoes before throwing them so no one gets hurt. You are not allowed to throw tomatoes directly at buildings. No bags or valuables are allowed, and it's advisable to wear googles or glasses as tomato juice is acidic so it can sting. But above all, the main aim is to have fun, so just enjoy!

AFTERMATH

After around an hour of squishing, throwing, and ducking, the event comes to a close. At this point, no more tomatoes can be thrown, and fire trucks move in to wash everything down. Of course, this is meant to clean the streets, but the crowd loves to get involved too! Some people go to the Los Peñones pool or Buñol River to wash off. It's likely your clothes won't survive the tomato stains, but they make a great souvenir.

MORE FUN!

And after all that frenetic energy and mess, it's time to enjoy something a little more sedate. The port city of Valencia, around 25 miles (40 km) to the east of Buñol, boasts numerous historic buildings, including the 14th-century Torres de Serranos (below), a modern interactive science museum, the City of Arts and Sciences (Ciudad de las Artes y las Ciencias), plus plenty of sandy beaches and water sports.

Watch volcanoes explode...

...AT STROMBOLI AND MOUNT ETNA

With its sandy beaches and ice cream aplenty, the island of Sicily in Italy may be best known as a summer vacation destination. But volcano tourism is another reason people visit... The active volcano Mount Etna dominates the east of the island and Stromboli sits just off the north coast. Both erupt regularly, and while lava-chasing may seem exciting to some, it can be pretty risky – after all, a huge eruption may be just around the corner. So get ready to see the power of planet Earth in action, but also remember to take all the advised safety precautions!

DANGER IN YOUR OWN BACKYARD

Believe it or not, around 500 people call Stromboli home. While the residents run the risk of having to evacuate at a moment's notice should a huge eruption occur, they do benefit from fertile lands and stunning scenery. Why not hire a bike and cycle around Stromboli like the locals? Or head out on a boat and take in the view from the water if you'd prefer a safer option!

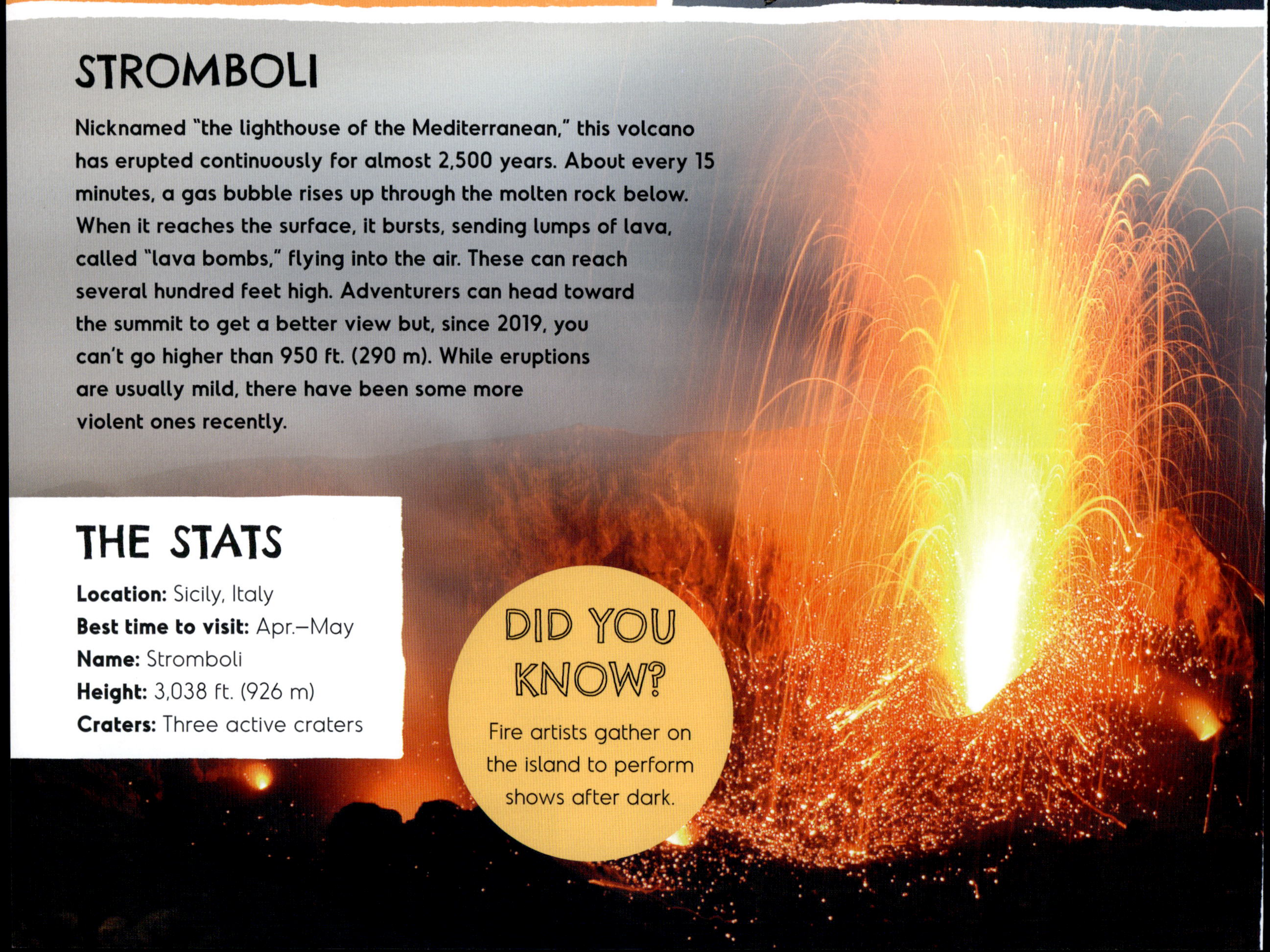

STROMBOLI

Nicknamed "the lighthouse of the Mediterranean," this volcano has erupted continuously for almost 2,500 years. About every 15 minutes, a gas bubble rises up through the molten rock below. When it reaches the surface, it bursts, sending lumps of lava, called "lava bombs," flying into the air. These can reach several hundred feet high. Adventurers can head toward the summit to get a better view but, since 2019, you can't go higher than 950 ft. (290 m). While eruptions are usually mild, there have been some more violent ones recently.

THE STATS

Location: Sicily, Italy
Best time to visit: Apr.–May
Name: Stromboli
Height: 3,038 ft. (926 m)
Craters: Three active craters

DID YOU KNOW?

Fire artists gather on the island to perform shows after dark.

MOUNT ETNA

Also called Mongibello, Mount Etna is the highest active volcano in Europe, and has been active for the past 2.6 million years! To get up the volcano, you first take a cable car (below) to 8,200 ft. (2,500 m). You can then either join a guided hike of the area or take a ride in a 4x4. There are four active craters to spot, but you should also look out for steam vents and pathways created by ancient lava flows.

THE STATS

Location: Sicily, Italy
Best time to visit: May and Sept.
Name: Mount Etna
Height: 10,900 ft. (3,320 m)
Craters: Four active craters

DID YOU KNOW?

Like Stromboli, people live near Mount Etna. The city of Catania at the foot of the volcano is actually Sicily's second-largest city.

CREATING CAVES

Once you've finished exploring Etna aboveground, you can also visit some of its 200 caves. These underground caverns were created by flowing lava – the outer crust would have cooled and solidified to form a roof, while the lava inside maintained its heat and continued to flow. The Ice Cave is located at 6,690 ft. (2,040 m) high and is part of the most southerly glacier in Europe. The Raspberry Cave (left) has interesting holes in the roof, allowing light to enter and make magical scenes. You can also visit the Three-Levels Cave, which has three overlapping and connecting galleries and is the longest of the caves, at 3,770 ft. (1,150 m).

Tandem paraglide...

...OVER A

FAIRY-TALE CASTLE

Get ready for some epic views as you glide over alpine mountains, ice-blue lakes, and the palace that inspired Walt Disney. Welcome to Bavaria, Germany – the home of the iconic Neuschwanstein Castle. Built by King Ludwig II in the mid-1800s, this dream home in the mountains was inspired by his love of art and poetry. You can buy a ticket to tour the inside of the castle, but viewing it from the sky is an even more magical way to take in its beauty.

A NEW SWAN

King Ludwig II was a dreamer. His childhood was filled with German fairy tales and theater, and he had a passion for composer Richard Wagner's operas. He wanted his castle's interior to be decorated with elaborate scenes from German legends, and made sure it included a sumptuous Throne Hall, a Singers' Hall featuring painted signs of the zodiac, and even a grotto (below), complete with waterfall and artificial stalactites! Look for swan emblems throughout – Ludwig identified with the legend of the swan knight Lohengrin, and the castle's name means "new swan stone."

CREATIVE MASTERPIECE

King Ludwig wanted his private palace to be constructed in the same style as the knights' castles of medieval Germany. However, his vision of a dreamy retirement home in the hills didn't quite work out... The construction work that he hoped would take three years ended up taking over seventeen. Unfortunately, he died in 1886 just a few weeks before the masterpiece was fully complete.

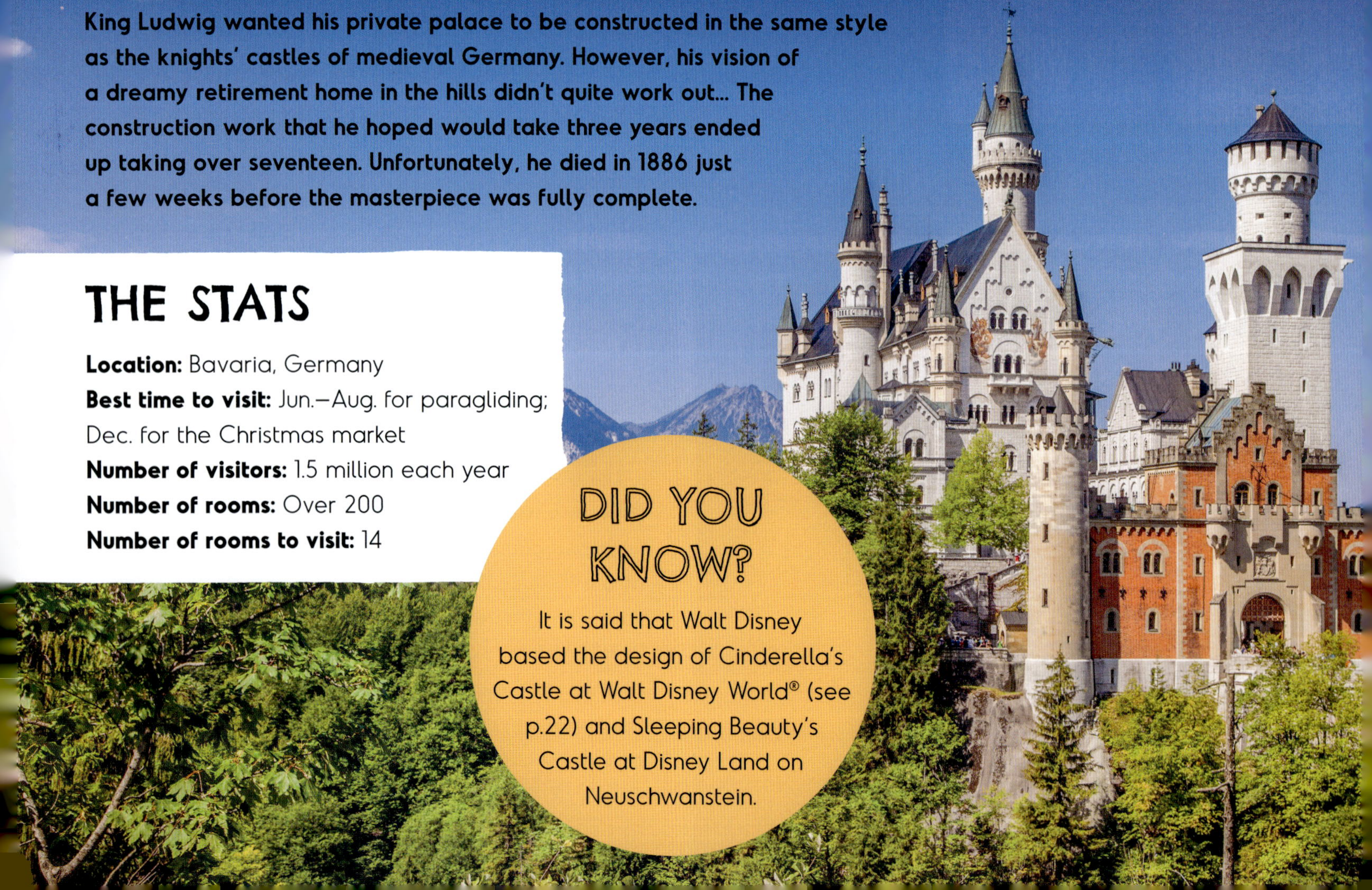

THE STATS

Location: Bavaria, Germany
Best time to visit: Jun.–Aug. for paragliding; Dec. for the Christmas market
Number of visitors: 1.5 million each year
Number of rooms: Over 200
Number of rooms to visit: 14

DID YOU KNOW?

It is said that Walt Disney based the design of Cinderella's Castle at Walt Disney World® (see p.22) and Sleeping Beauty's Castle at Disney Land on Neuschwanstein.

PARAGLIDE WITH A PARTNER

Admire Neuschwanstein from above as you paraglide down Tegelberg Mountain. This free-flying adventure sport is easier to experience than you may think – professional tandem flights operate in the area, allowing you to simply clip onto a pro pilot and take to the skies! Paragliding uses just a single wing – the air is your engine. Gaze down at emerald lakes and the River Lech as well as the lush mountains that surround the castle. When you're ready to land, your pilot will fly figures of eight to reduce height, then stall the wing so you drop down and land... on your feet, hopefully!

MYSTERIOUS ENDING

Unfortunately, Ludwig's fantasy project of building the castle was not so dreamy for the state of Bavaria, which was paying for it all. The costs nearly bankrupted Bavaria, and ministers declared Ludwig insane. They sent him to stay at the Castle of Berg but, the day after his imprisonment there, Ludwig was found dead in nearby Lake Starnberg. It's said Ludwig was a strong swimmer and the water wasn't deep, so the case is still open in this story of the fairy-tale king's demise.

MORE FUN!

Winter is a special time of year to visit Bavaria, and its many Christmas markets only add to the magic of the region. Fussen is the closest town to Neuschwanstein Castle, and is famous for its high-quality violin-making. At Christmas time, a small market of craft and food stalls is set up in the courtyard of the St. Mang monastery, and choirs from the region come to perform. It's a truly magical setting in the view of the castle, and twinkling Christmas lights make the historical town center glow.

Next, it's...

...AWE-INSPIRING

AFRICA

Africa is all about the animal experiences... It's the second-biggest continent in the world, so there's plenty of space for them to roam free. Of course, you're going to want to see the famous Great Migration as well as classic savannah species, such as lions, giraffes, and elephants. But you can also try your luck looking for sharks in South Africa and lemurs in Madagascar. Aside from the animals, there are zip wires to whizz down, giant waterfalls and ancient pyramids to be awestruck by, and sprawling markets to do some serious souvenir shopping in. Whatever treasures you hope to find, your African experience will be one to remember.

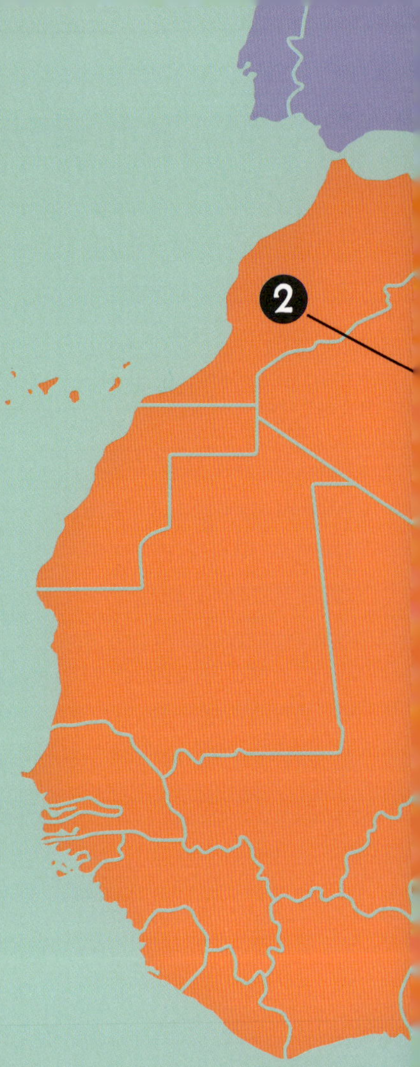

PLACES TO VISIT:

1. **The Great Pyramids of Giza**, Cairo, Egypt, p.88
2. **Marrakesh**, Morocco, p.92
3. **Luxor**, Egypt, p.89
4. **Dallol Hydrothermal Field**, Danakil Depression, Ethiopia, p.94
5. **Erta Ale Volcano**, Ethiopia, p.95
6. **Masai Mara National Reserve**, Kenya, p.90
7. **Serengeti National Park**, Tanzania, p.90
8. **Victoria Falls**, Mosi-oa-Tunya National Park, Zambia, p.102
9. **Forest of Knives**, Tsingy de Bemaraha National Park, Madagascar, p.98
10. **Avenue of the Baobabs**, Madagascar, p.99
11. **Okavango Delta**, Botswana, p.96
12. **Cape Town**, South Africa, p.101
13. **Tsitsikamma National Park**, South Africa, p.104
14. **Shark Alley**, South Africa, p.100

WET AND DRY

Look out for the River Nile running down the east side Africa. It's the longest river in the world and gives life to what is otherwise a very dry region. The Sahara Desert in the north of the continent is the largest hot desert in the world.

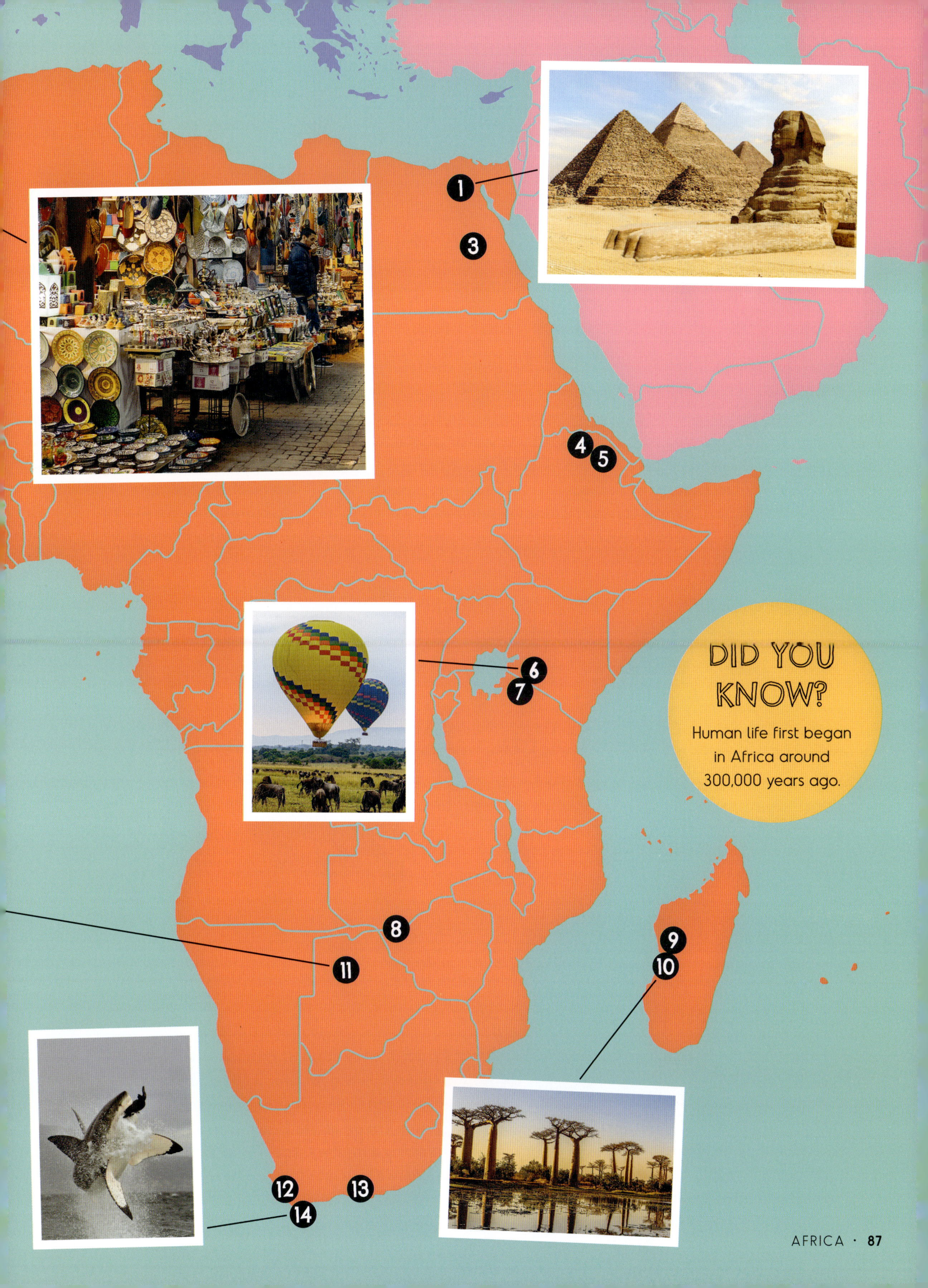

DID YOU KNOW?

Human life first began in Africa around 300,000 years ago.

Discover the secrets of the pharaohs...

...AT THE GREAT PYRAMIDS

THE STATS

Location: Giza, Cairo, Egypt
Best time to visit: Nov.–Feb. for cooler weather
Built: c.2550–2490 BCE
First pharaoh: Menes, c.3000 BCE
Last pharaoh: Cleopatra, 69–30 BCE

The three huge pyramids and the complex around them at Giza are some of the most iconic structures in the world. Seeing them up close will take your breath away. But what lies within them? You've probably heard stories of endless treasure hidden inside... could there still be more to find? Join a tour and you'll learn all about the ancient Egyptians who built these giant stone tombs as final resting places for their kings and queens, the pharaohs.

DID YOU KNOW?

The Great Sphinx has a lion's body and a human head, and was built to guard the pyramid tombs of the pharaohs. Look closely and you'll see it's missing its nose!

OUTSIDE AND INSIDE

The biggest of the three pyramids, known as the Great Pyramid, is the tomb of Pharaoh Khufu (above right). Originally 481 ft. (146.6 m) high, it was the world's tallest building for over 3,800 years. It sits alongside the pyramids of the pharaohs Khafre (above center) and Menkaure (above left) as well as some smaller pyramids built for the pharaohs' queens. A huge statue known as the Great Sphinx stands nearby. Just how the pyramids were built, over 4,500 years ago, remains a mystery. You can go inside all three. In the Great Pyramid, a deep, narrow tunnel leads you to where Khufu was laid to rest with his riches. Sadly, the tomb was robbed soon after the burial, so no treasure remains today.

GRAND EGYPTIAN MUSEUM

The pyramids themselves are now empty of treasure, and so you'll need to visit the nearby museum to see any artifacts. Although burial sites were often looted, over 100,000 items have been collected to showcase the amazing Ancient Egyptian civilization. The most famous pieces come from Tutankhamun's tomb. He wasn't buried at Giza, but at another important site called the Valley of the Kings. The tombs here were much less imposing than the pyramids, and amazingly, Tut's tomb, cut into the ground, was never found by robbers. Instead it was discovered by an archaeologist called Howard Carter in 1922.

TUT'S TREASURE

Imagine Howard Carter's face when he discovered Tutankhamun's tomb hidden in the ground, filled with over 5,000 objects for use in the afterlife. A mannequin would have helped the young pharaoh choose what to wear each day – looking beautiful was a sign of holiness and a way of pleasing the gods of the afterlife. There were also board games to play, a "golden throne" to sit on, and the star attraction of the museum, Tut's golden coffin and jeweled death mask (right).

MORE FUN!

After spending time at Giza, take a cruise along the River Nile to Luxor around 400 miles (650 km) away. Your days on the water will give you a feel for how important the Nile is to Egyptians – its annual floods provide fertile soil for farming in what would otherwise be a desert. Luxor was originally known as Thebes and was Ancient Egypt's capital for 1,500 years. Here you'll find a great open-air museum of temples, tombs, and monuments created by many powerful pharaohs. The Valley of the Kings is also nearby, so perhaps you'll recreate Howard Carter's famous discovery!

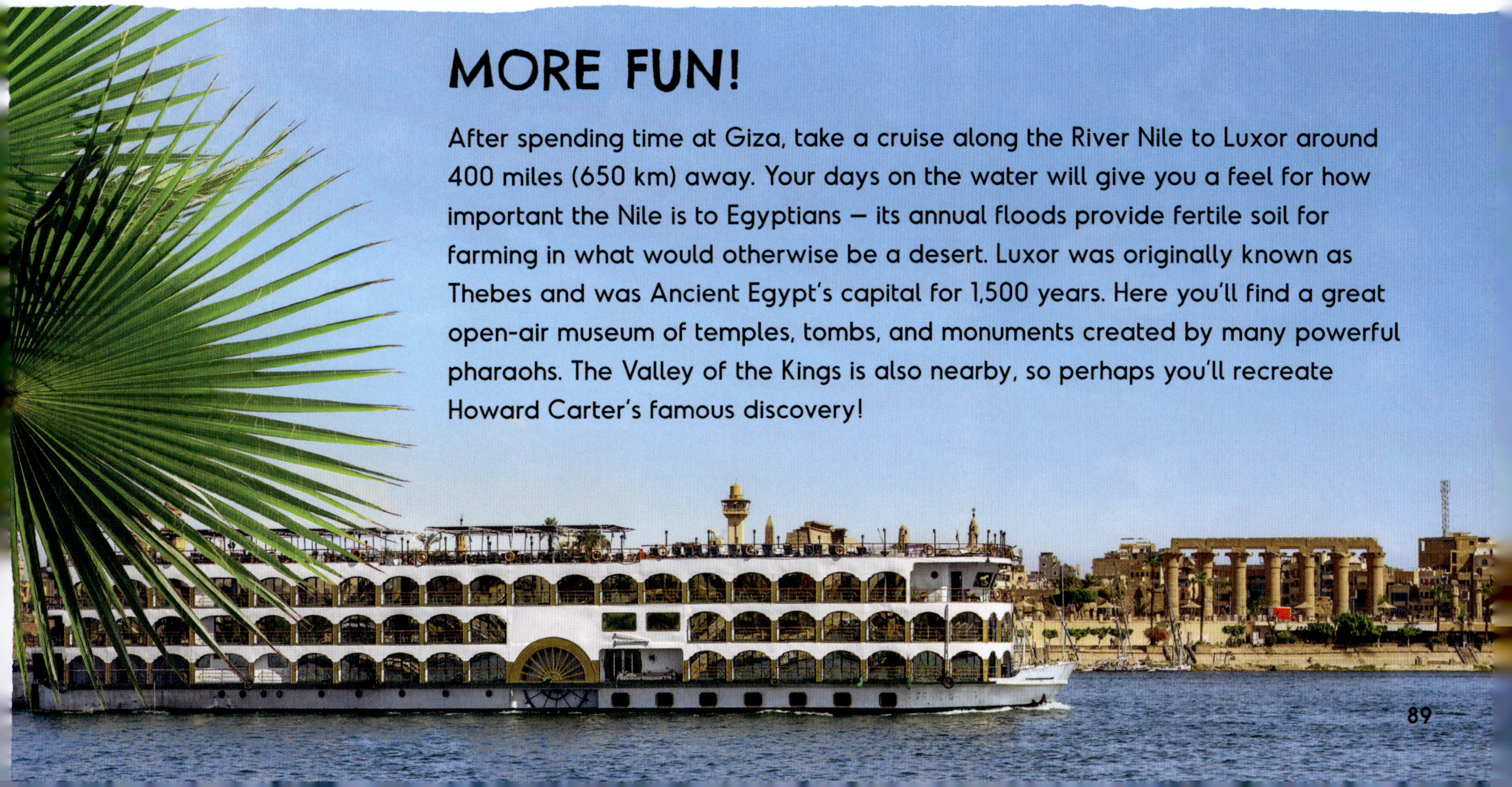

Witness the world's biggest wildlife spectacle...

...AT THE GREAT MIGRATION

You're on safari in Africa, the ground starts to vibrate, and you hear a distant thundering noise, getting louder and louder. This can only mean one thing – 1.5 million wildebeest and 200,000 zebras and gazelles are about to make their presence known! Travel to Kenya in the summer months and see firsthand one of the most amazing wildlife experiences, the Great Migration.

MAJOR MOVEMENT

The Great Migration is the continuous annual movement of animals from one area of a vast savannah (grassland) to another, in a never-ending search to find fresh grass and water. After giving birth in the early months of the year in the Serengeti, Tanzania, the wildebeest know that the plains here will dry up in the summertime. So, in late spring, they set off on an epic expedition north to the Masai Mara, Kenya, in search of better grazing. The herds are mostly made up of wildebeest, but other herbivores make the journey too, including zebras and gazelles.

THE STATS

Location: Serengeti National Park, Tanzania, and Masai Mara National Reserve, Kenya
Best time to visit: Jul.–Sept. for the river crossings
Best place to view: The Mara River
No. of animals: More than 2 million large herbivores
Migration distance: Over 1,000 miles (1,600 km)

RISKY BUSINESS

As the giant herds move, they are followed by a small army of predators, such as lions, cheetahs, and hyenas. Those successful in making it past the claws and jaws of the savannah's hungry meat-eaters soon face another danger – and this one is lurking underwater... To reach the Masai Mara the animals must cross the Mara River (bottom left), home to hundreds of hungry crocodiles. The nervous wildebeest will often wait for hours, even days, before choosing their moment to cross. But the crocodiles have been waiting all year for this moment, and they don't miss an opportunity to fill their bellies!

CHOOSE YOUR RIDE

Your best chance at spotting the migrating animals – and their many predators – is aboard a safari jeep. These vehicles are built for the terrain and usually have open sides, or an open roof hatch, to give good views. To upgrade your experience, there's also the option of a hot-air balloon ride for a bird's-eye view over this incredible landscape (right). And if money is no object, you could take a scenic flight in a small plane. The animals are used to the noise at the Musiara airstrip in the Masai Mara, so you may end up landing next to an ostrich or an elephant!

MOVING ON

By September, the Masai Mara plains are full of wildlife. But by late October, the weather starts changing again, and the wildebeest, gazelles, and zebras begin to make their way back to the Serengeti to have their young. The soil there is rich in nutrients following volcanic eruptions millions of years ago, so mothers eating this grass make richer milk for their calves. As the months pass, the whole process starts again, and the next generation is introduced to this epic migration route.

Go on the ultimate souvenir shopping trip...

...IN THE SOUKS OF MARRAKESH

Grab your shopping bags and get ready to explore one of North Africa's most vibrant cities. Marrakesh is a hotbed of Moroccan culture and an ancient trading hub that is home to the famous souks (markets) where you can shop until you drop. The center of this city is called the medina and it's here, inside its red walls, that your senses will be awoken with an explosion of sights, smells, and tastes. In fact, the whole country could be described as a sensory delight, with mountains, deserts, beaches, and even skiing available!

SHOPPING IN THE SOUKS

The souks in Marrakesh offer a shopping experience like no other. Colorful clothing, shiny trinkets, spices, and candy are all laid out ready to be scooped up – it'll be hard to resist at least one purchase! The stalls are arranged in a labyrinth of lanes, and each area specializes in different products, such as items handmade by Berber tribes living in the Moroccan mountains and deserts. For a bit of fun, try your hand at haggling. This is where you engage in friendly conversation with a seller in order to negotiate a price!

THE STATS

Location: Medina old town, Marrakesh, Morocco
Best time to visit: Year-round, in the morning when it's cooler and quieter
Total number of stalls: Around 3,000
Items sold: Spices, housewares, kitchen items, lamps, leather goods, clothing, trinkets
Opening hours: Roughly 9:00 a.m.–9:00 p.m.

FOOD FOR THOUGHT

Marrakesh is one of the top destinations in the world for a street food experience. At night, the Jemaa el-Fnaa square in the medina comes to life with stallholders setting up their kitchens and smells filling the air. This square has been used as a food market since the 11th century, and local favorites include tagine – meat with spices, slow-cooked in a dome-shaped clay pot – served with couscous.

EVENING ENTERTAINMENT

Every night, the Jemaa el-Fnaa also hosts a variety of entertainers. Musicians, comedians, acrobats, and artists all put on a show for visitors. There's a lot of hustle and bustle to contend with, so if you prefer, head to one of the rooftop cafés surrounding the medina to watch it all unfold from above.

JARDIN MAJORELLE

After the excitement of the medina, you'll probably want a taste of tranquillity, so head to the gardens of the Jardin Majorelle (below) to experience nature and art together. Created in 1924, the gardens were later taken over by French fashion designer Yves Saint Laurent in 1980. You'll find native Moroccan plants here as well as species from five different continents, but the electric-blue architecture and museum of traditional arts are also what bring visitors to the site.

DID YOU KNOW?

You can do it all in Morocco: skiing in the Atlas Mountains one day, sand-boarding in the Sahara the next, then relaxing on a beach the day after that.

MORE FUN!

Need a break from the buzz? It doesn't take long to head out of Marrakesh and into the desert, where you can ride a camel and experience some traditional Berber hospitality. The camels found in Morocco are dromedaries, which means they only have one hump. The ride will feel a bit like you're in a rocking chair – if you're young, then make sure you have an adult with you to hold on to!

Marvel at an alien landscape...

...IN THE DALLOL HYDROTHERMAL FIELD

Strange colors and patterns make this area in Ethiopia seem like something from outer space. With the pungent scent of smelly toxic gases in the air, this is not a tourist attraction many rush to see. So why is this spot, the Dallol Hydrothermal Field, formed somewhere between 5.3 and 23 million years ago, something you must experience? Because it is truly unique, geologically fascinating – and there's nothing else quite like it on Earth!

CRAZY COLORS

Acidic water containing chemicals from inside the volcano rises to the surface, creating pools of vibrant colors! Toxic yellows, eerie greens, brilliant browns, and burnt oranges might not seem natural, but they are. Created by different variations of minerals containing iron and sulphur, these chemicals are what cause the stench in the air... so it's best to not breathe in too deeply.

RECORD BREAKER

The site of the Dallol Hydrothermal Field is just 0.4 sq. miles (1 sq km) in size, and it's one of the lowest points in Africa. It's found next to Ethiopia's most active volcano, Erta Ale, and so it is also one of the hottest year-round places on Earth. It's this combination of heat, plus the volcanic activity beneath the ground, that results in such an unusual appearance on the surface.

SALT MOUNDS

Pools of water are heated by the high surface temperatures, which causes much of the liquid to evaporate, leaving behind bizarre cones of salt. The remaining water has a very high acid content, so while you might feel like cooling down, don't be tempted to go for a dip! If that doesn't sound hazardous enough, beware also of steaming fumaroles and mini-geysers that shoot boiling hot water into the air. The landscape is ever-changing due to all this hydrothermal activity – so watch your step!

THE STATS

Location: Danakil Depression, Ethiopia
Best time to visit: Sept.–May
Size: 0.4 sq. miles (1 sq km)
Altitude: 330 ft. (100 m) below sea level
Water temperature: Some pools over 194°F (90°C)
Air temperature: Up to 122°F (50°C)

DID YOU KNOW?

The area is often referred to as the "Gateway to Hell" due to its high temperatures and hostile environment.

SMOKING MOUNTAIN

The nearby volcano Erta Ale is what's known as a shield volcano. It is 2,011 ft. (613 m) high and has a bubbling lava lake at the summit. When pressure inside the volcano rises, it causes the lava in the lake to boil over and flow down the side of the volcano and harden, creating the gently sloping sides that classify it as a shield volcano. The name Erta Ale means "smoking mountain" in the local Afar language.

Squelch through the world's greatest swamp...

...IN THE OKAVANGO DELTA

Surrounded by the dry Kalahari Desert, the Okavango Delta is a large, flat wetland in northern Botswana that creates a lush habitat for all kinds of animals. This vast area of grassy plains is flooded annually by the Okavango River, so animals from all over come here to eat, drink, and be merry. Embark on a water safari to see some of the most spectacular wildlife Africa has to offer.

HUGE HIPPOS

The Okavango's most famous residents are its hippos. These huge beasts can weigh up to 3,300 lb. (1,500 kg) and measure 13 ft. (4 m) from nose to tail. Despite their love of the water, they are not very good swimmers, and instead move around by walking or jumping. Hippos have their eyes, ears, and nostrils on the top of their heads so they can still use their senses when the rest of their body is underwater. Watch out for their powerful jaws and long teeth, and keep your distance – an angry hippo is a dangerous animal!

CRUISE THE WATERS

Seeing wildlife from the water is what an Okavango experience is all about. Ride in a traditional *mokoro*, a dugout canoe, for a serene journey. It'll allow you to cruise through narrow waterways, cross huge lagoons, and even reach areas where the savannah grasslands run right up to the banks of the lush wetlands. There's no engine to disturb the animals, so keep your camera at the ready to snap the perfect photo.

THE STATS

Location: Botswana
Best time to visit: Mar.–Sept. for high water
Size: 6,000 sq. miles (16,000 sq km)
Water depth: 3–65 ft. (1–20 m)
Length of river: 680 miles (1,100 km)
Age of the delta: 2.5 million years

DID YOU KNOW?

The wetlands are home to an incredible number of different animals: over 400 species of birds, 155 species of reptiles, and 160 species of mammals.

BEAUTIFUL BIRDS

Botswana is one of Africa's top bird-watching destinations because so many migratory birds visit the area. Birdsong can be heard wherever you go, and flashes of color will catch your eye at each turn. But, most interestingly, by watching the birds you can tell when the floodwaters are coming. The wattled cranes, yellow-billed storks, herons, and egrets all move into the Okavango Delta to feed on the insects and small creatures that try to escape from other areas as the waters rise.

Wattled crane

Heron

Yellow-billed stork

Egret

MAJESTIC ELEPHANTS

Botswana has Africa's largest population of elephants, estimated at around 130,000 individuals. They eat so much vegetation every day that their influence shapes the flow of the water around the delta – they keep channels open and form new ones as they fell big trees or open up dams. The elephants' size means they can easily wade through passages of water that other animals can't cross. If you stay at the Abu Camp in the delta, you may have the opportunity to walk alongside the famous Abu elephant herd.

MORE FUN!

For an even more extraordinary experience, you could stay overnight in the delta at a special water camp. These bases by the water will not only give you the chance to take daily safaris, but will also give you more opportunities to see rare animals like the sitatunga antelope, whose wide hooves are adapted for walking through swampland, and the wattled crane, often spotted in breeding pairs. It may take you a while to get to sleep, though. Being right by the water means you'll also come close to the African bullfrog that will happily put on a concert of croaks for you!

Look for lemurs...

...HIDING IN MADAGASCAR'S FOREST OF KNIVES

On the west side of the island of Madagascar, there's a place unlike any other. Towering stone needles appear to rise up out of the surrounding forest, and caves provide secret hiding places for animals not found anywhere else in the world. Exploring Tsingy de Bemaraha, also known as the "Forest of Knives," is not for the fainthearted. But the rewards are worth it – you'll be able to observe the many different species of lemurs and rare birds that call this unusual habitat home.

Ring-tailed lemur

Von der Decken's sifaka lemur

HIDDEN HISTORY

The adventure begins before you even set foot in the Forest of Knives. To get to this unique national park, you'll first travel to the remote village of Bekopaka. The roads are rough and there's a river to cross, which gets so high in the wet season it's nearly impossible to get here. But once you do make it to the "forest," it'll take your breath away. Around 200 million years ago the limestone seabed rose, creating a plateau, which was then eroded over the years by rainwater into peaks, canyons, and caves. The erosion happened both horizontally and vertically, which is how the distinct needle shapes were created.

DID YOU KNOW?

World Lemur Day is the last Friday in October.

LOVELY LEMURS

Lemurs are primates – they belong to the same family as monkeys and apes (and us!). Madagascar is the only place in the world where lemurs live in the wild, and the lack of predators on the island means they have evolved into various shapes and sizes. Most people are familiar with the iconic ring-tailed lemur (top left), but there are over 100 other species. The smallest is the Madame Berthe's mouse lemur, which weighs just 1 oz. (30 g). The largest is the indri, which, fully grown is about the same size as a toddler and weighs about 21 lb. (9.5 kg)!

TREE LOSS

Sadly, lemurs are facing extinction due to the loss of their forest home. Trees are not being planted quick enough to replace those cut down. As with much conservation these days, education is a key part to helping locals find ways to protect these animals while also farming the land.

ROPE WALKS

Tsingy de Bemaraha roughly means "where one cannot walk barefoot," and it's certainly true of this spiky UNESCO World Heritage Site. Since the installation of aerial suspension bridges, it's now much easier to explore the area, and tour guides offer trail walks and rope climbs. If you choose to join the Anjohimanintsy Trail, you'll also camp in the park overnight. After exploring the limestone peaks throughout the day, you'll descend into the caverns to set up your tent. You might even sleep in the same space as a lemur!

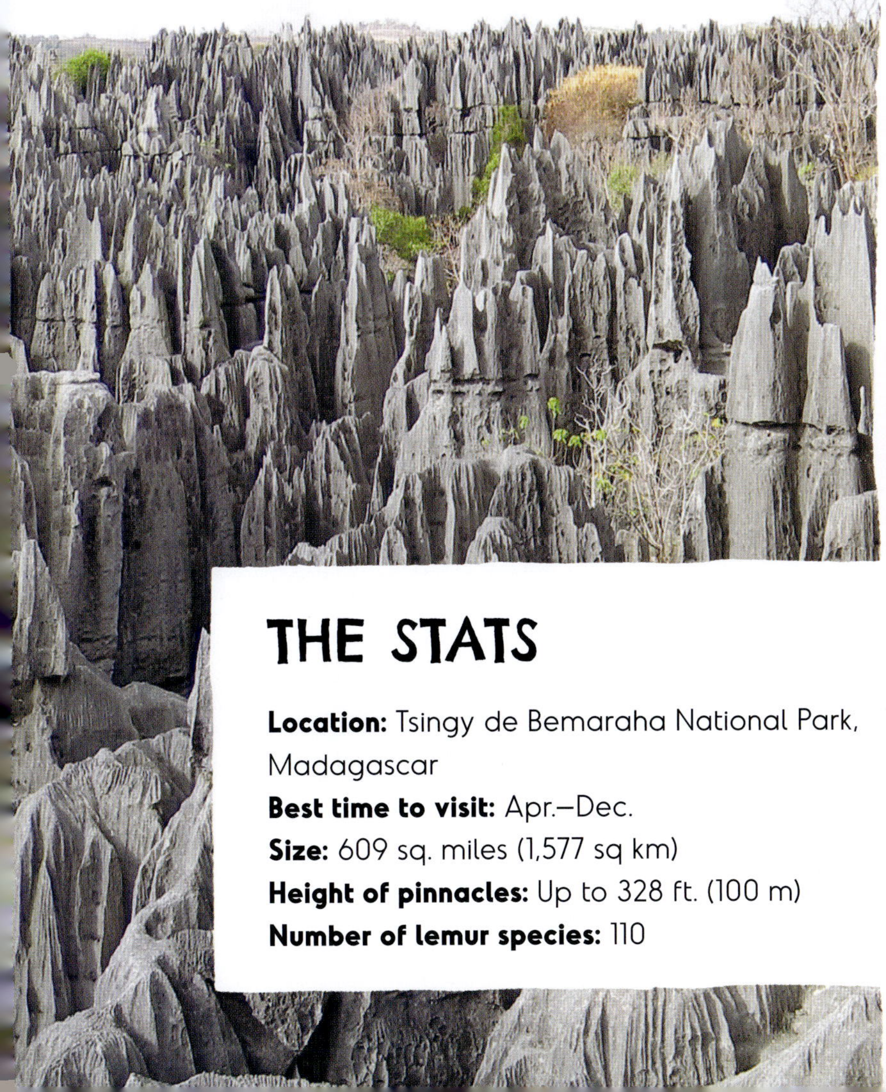

THE STATS

Location: Tsingy de Bemaraha National Park, Madagascar
Best time to visit: Apr.–Dec.
Size: 609 sq. miles (1,577 sq km)
Height of pinnacles: Up to 328 ft. (100 m)
Number of lemur species: 110

MORE FUN!

Madagascar is a land of many different experiences. After your adventure in the Forest of Knives, why not head to the golden shores to discover a deserted beach, or trek into the rainforest to look for more unique wildlife such as the leaf-tailed gecko? You could even swim with whale sharks if you head out to the small island of Nosy Be. And no visit to Madagascar would be complete without seeing the huge baobab trees in the famous Avenue of the Baobabs (above).

Study great whites...

...DOWN IN SHARK ALLEY

Shark Alley – a strip of sea between Dyer Island and Geyser Rock – off the coast of South Africa near Cape Town is a shark playground! Hundreds of them gather to feed on the 60,000-strong colony of Cape fur seals that live and breed here. So, if you're brave enough, this is the place to come to get closer than you could ever imagine to the most ferocious shark of all, the great white.

MEALTIME

Seeing a top predator in the wild is a thrilling experience. You may get lucky and even see a great white shark hunt and catch prey. Great whites have limited eyesight, and mainly hunt using their ability to detect the electric signals given off by other animals. Once they sense prey, they'll often position themselves underneath it before swimming up at high speed to stun it. If the prey is near the surface (seal pups are the dish of choice!), the shark will grab it between its jaws then dramatically lunge out of the water. What a sight!

DID YOU KNOW?

The minimum age to cage dive is usually ten, but some companies allow eight-year-olds to join the experience.

CAGE DIVE

Coming face-to-face with a great white shark is a dream for some and a nightmare for others. But, as you'll join a tour boat for this experience, you'll be using a key bit of safety equipment... a solid metal cage. Cage-diving is a way for tourists and scientists alike to be able to study these amazing predators up close without risk.

HOW IT WORKS

A cage is lowered into the water next to the boat, with the top of the cage remaining above the surface. Wearing a wetsuit, you'll ease yourself into the cage before bait (food) is dropped into the sea nearby. When the sharks appear, you're sure to get an adrenaline rush like no other! And if you're brave enough to keep your eyes open, you can help collect valuable data by recording the sharks' size, markings, and behavior.

SHARK RESEARCH

Sadly, the number of great white sharks off the coast of South Africa is in decline, and today there are believed to be only around 300–500 individuals left. But research can help us learn more about these animals. The first step is to identify individuals using the shape and markings of their dorsal fins in order to build up a database. Teams can then study each shark's movement patterns and growth rates every time they take a cage-diving boat out onto the water. When you're older, you could even apply to volunteer with one of the conservation programs!

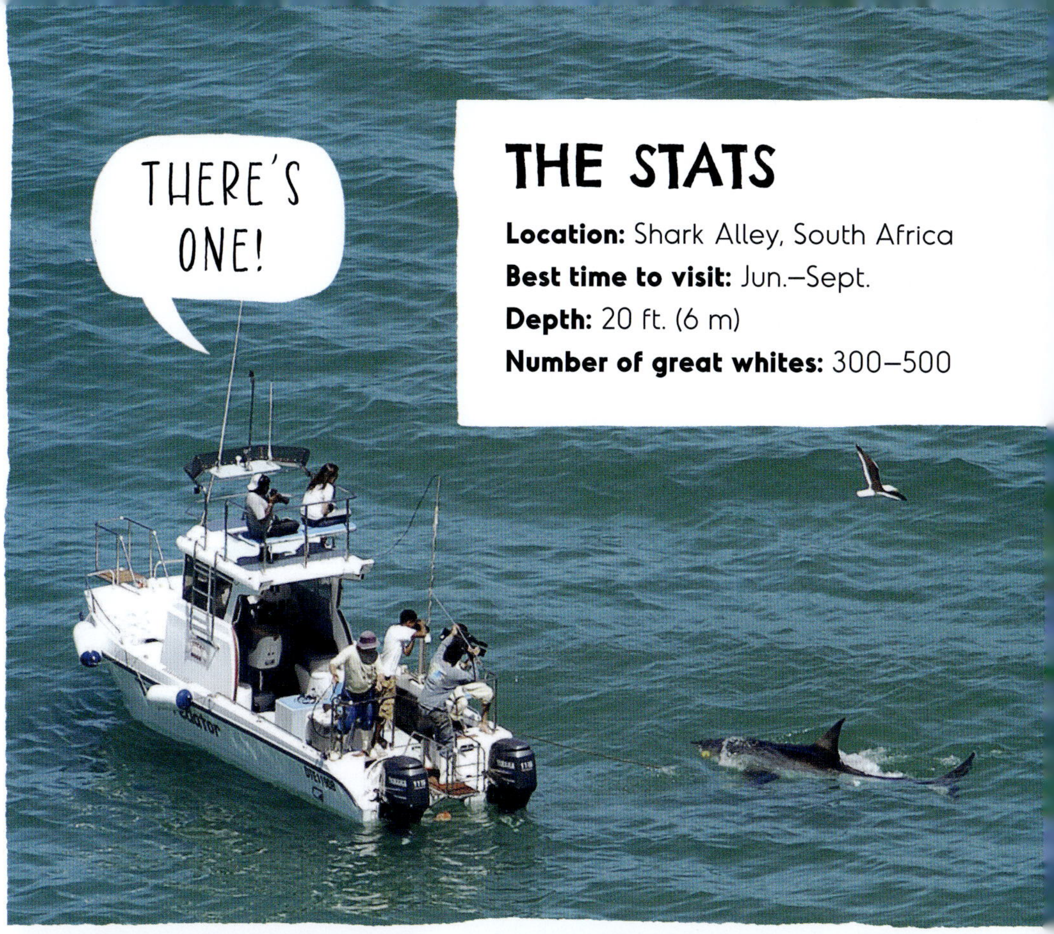

THE STATS

Location: Shark Alley, South Africa
Best time to visit: Jun.–Sept.
Depth: 20 ft. (6 m)
Number of great whites: 300–500

MARINE BIG FIVE

Just like the "Big Five" land animals of Africa that everyone wants to see, there is a "Marine Big Five" too. Of these, the African penguin is probably the easiest to spot, as a large colony lives in the marine protected area of Table Mountain National Park at Boulders Bay.

1. Great white shark, 2. Southern right whale, 3. Bottlenose dolphin, 4. Cape fur seal, 5. African penguin.

These can all be spotted along the Gansbaai coastline, just south of Cape Town.

MORE FUN!

One of South Africa's three capital cities (along with Pretoria and Bloemfontein), nearby Cape Town sits in the shadow of the impressive Table Mountain. It's named for its flat top, which is sometimes covered with a layer of cloud, fondly called the "tablecloth." Take a cable car to the top for amazing views (when it's not cloudy!).

Bombard your senses and get soaked...

...AT

VICTORIA FALLS

For an awe-inspiring sight that showcases the power of nature, head to Victoria Falls on the Zambezi River in Zambia. This is officially the largest curtain of falling water in the world, and the clouds of mist and (sometimes double!) rainbows that surround it make it look like it belongs in a magical land. There are many ways to experience the falls, and with so many changes throughout the year, this is one landmark you may wish to visit again and again – and again!

THE STATS

Location: Mosi-oa-Tunya National Park, Zambia

Best time to visit: Mar.–Apr. for high water; Sept. for low water

Falls height: 354 ft. (108 m)

Falls width: 5,603 ft. (1,708 m)

PEAK FLOW

Visit in March or April and you'll get to see the gentle Zambezi River transform into a torrent of white water tumbling over the cliff edge. Walk 130 ft. (40 m) across the Knife Edge Bridge to get a closer look at the Eastern Cataract section of the falls, as well as the Main Falls and the Boiling Pot, where the river swirls down the Batoka Gorge. When the waters reach their peak level, it really is an assault on the senses – you'll be able to taste, touch, hear, and smell the water spray swirling in the air around you. It goes without saying that wearing rain gear is a must.

DID YOU KNOW?

The Zambezi River forms the border between Zimbabwe and Zambia, and you'll get a different view of the falls depending on which country you're in. Cross the Victoria Falls Bridge to travel between the two countries.

DRY DISCOVERY

You may think that the falls will be less dramatic in the dry season – after all, as the flow of the river slows, less water gushes over the edge. But the drop in water level and ferocity means it's possible to get to Livingstone Island at the lip of the falls. Book a ticket to visit this unique island and a boat will transport you there. Your tour guide will tell you about the history of the falls. They were named by British explorer David Livingstone, after the then British monarch, Queen Victoria, but are known in the local Sotho language as Mosi-oa-Tunya (the smoke that thunders).

DEVIL'S POOL

Another bonus of the dry season is that you have the chance to swim up to the very tip of the falls! It may sound impossible, but a rock pool on the edge of Livingstone Island provides a safe space for those aged 12 and over to take a dip. It's away from the strong currents, so you can sit back in the water and relax – and if you're brave enough – dare to look over the edge!

ABOVE AND BELOW

Other ways to experience these magnificent falls are from above and below. Take a helicopter flight for spectacular aerial views of Victoria Falls and the raging waters below (right). You can even swim in the rock pools right underneath the waterfall in the dry season!

MORE FUN!

The Zambezi River is one of the best white water rafting locations in the world. The river flows over the cliff face and drops down into the Batoka Gorge – a deep channel cut into the rock over millions of years by the gushing water. It is here, beneath the falls, that the challenge of a lifetime awaits those who are brave enough – and old enough (you have to be 15) – to jump in a boat and tackle sections of water terrifyingly known as Oblivion and Terminator. It's also great fun to watch if you're not quite ready to take it on in person yet!

Give yourself a rush of adrenaline...

...ZIPPING THROUGH GIANT TREES

South Africa is a great place to try extreme sports, explore the countryside, and enjoy magical wildlife encounters – so why not do them all at once? At Tsitsikamma National Park and its surrounding area, you can try a range of heart-racing activities, including zip-lining and blackwater tubing – all while getting up close and personal with a variety of wildlife.

ZIP HERE, ZIP THERE

There are plenty of zip lines of various lengths and heights to choose from in the area. In the Addo Elephant National Park to the northeast of Tsitsikamma, there's the country's highest, fastest, and longest double zip line, allowing you to reach speeds of around 35 mph (60 km/h) with a buddy. You'll also find a superman-style zip line here – you can imagine how you'll look riding that one! Or head to Hottentots Holland Nature Reserve in the heart of a World Heritage Site near Cape Town (below), where you can zip over beautiful gorges, raging waterfalls, and sheer cliff faces.

ZIP TOUR

In the Tsitsikamma National Park forest, you'll find zip lines running through giant outeniqua yellowwood trees. Suspended up to 100 ft. (30 m) above the forest floor, you'll have 10 zip lines in total to try. Prepare to come face-to-face with a huge number of birds – and maybe a vervet monkey or two – as you glide through the canopy.

BUNGEE JUMP

To the west of the Tsitsikamma region you'll find the Bloukrans River. If you're 14 and over, this is the place to be if you want an additional adrenaline boost! The Bloukrans Bridge is home to the world's highest bridge bungee jump at 709 ft. (216 m). It also holds the record for the most bungee jumps in 24 hours: an amazing 107 jumps were completed by Scott Huntley in 2011.

THE STATS

Location: Tsitsikamma National Park, South Africa
Best time to visit: May and Nov.
Established: 1964
Zip tour length: 2–2.5 hours
Longest zip line: 298 ft. (91 m)
Animals to spot: Vervet monkeys, African crowned eagles

AQUA ADVENTURE

Once you're back on the ground, head to the Storms River Gorge for another adventure – blackwater tubing. Lie back and relax in your inflatable tube and let the gentle current transport you downstream. But beware – if it's been raining heavily, the calm waters will have transformed into white water rapids, and the experience will be more on the wild side! Whatever the weather, you'll be able to try stand-up paddle-boarding, tube down natural water slides made from rocks, and even jump from cliffs into pools of crystal-clear water.

MORE FUN!

Tsitsikamma National Park officially stretches for 3 miles (5km) out into the ocean as well. It's a Marine Protected Area and provides a perfect playground for two of its most famous visitors: the bottlenose dolphin and the Cape clawless otter. The otter is such a popular attraction that a five-day hiking tour along the coast has been named the "Otter Trail" in its honor. If you are aged 12 and above, you can take on the entire hike and cover 26 miles (42 km) of stunning coastline (below) during the day while resting your head in forest huts along the route at night.

Next it's...

...AMAZING

ASIA

In Asia, you can find the old and the new side by side. The traditional Hindu festival of Holi, celebrated across India, and the Lantern Festival in Vietnam contrast with the cutting-edge skyscrapers of Dubai and the futuristic gardens in Singapore. The natural world offers just as much intrigue in this vast continent. Do you dream of climbing the world's tallest peak, Mount Everest? Or perhaps a natural hot spring bath is more your sort of experience, just like the snow monkeys in Japan. And a tour of this continent wouldn't be complete without a visit to China, where you can visit a robot restaurant one day and watch pandas playing the next.

PLACES TO VISIT:

1. **Beijing**, China, p.120
2. **Fairy Chimneys**, Cappadocia, Turkey, p.108
3. **Jigokudani Monkey Park**, Japan, p.122
4. **Shiga Kogen Ski Resort**, Japan, p.123
5. **Tokyo**, Japan, p.124
6. **Terracotta Army Museum**, Xi'an, China, p.118
7. **Shanghai**, China, p.118
8. **Delhi**, India, p.112
9. **Mount Everest**, Nepal–China, p.116
10. **Kathmandu**, Nepal, p.117
11. **Dujiangyan Panda Reserve**, Sichuan, China, p.119
12. **Chengdu Panda Research Base**, Sichuan, China, p.119
13. **Taj Mahal**, Agra, Uttar Pradesh, India, p.115
14. **Burj Khalifa**, Dubai, United Arab Emirates, p.110
15. **Ranthambore National Park**, Rajasthan, India, p.114
16. **Foodom Robot Restaurant**, Guangzhou, China, p.119
17. **Old Quarter**, Hội An, Vietnam, p.128
18. **Gardens by the Bay**, Singapore, p.126

BIG AND BOLD

The largest continent by both size and population, Asia is big! India takes the top spot for most populated country, while Russia is the largest by area. With size comes extremes: towering mountains, dense jungles, and vibrant cities provide the setting for some incredible adventures.

DID YOU KNOW?

The Dead Sea, on Jordan's western border, is the lowest point on land – around 1,400 ft. (430 m) below sea level.

Take a hot-air balloon ride...

...OVER FAIRY CHIMNEYS

You're in Cappadocia, in Turkey, it's early in the morning – before sunrise – and the excitement is building. You hear a roar and the darkness is lit up by a burst of flames. Colorful balloons surround you, all waiting to ascend into the sky. Jump into your basket, take off, and float high above the magical landscape.

HIDDEN CITIES

Once your flight is well underway and the Sun has risen, look closely at the fairy chimneys and you will spot doorways in some. Believe it or not, these lead to a whole city of underground caves! These rock formations have been listed as a World Heritage Site by UNESCO, in part due to the cave-dwellings hidden beneath them.

FAIRY CHIMNEYS

As your balloon rises, you'll spot otherworldly pillars rising out of the ground. These unique formations were produced by volcanic explosions around 14 million years ago. The hot lava spread across the land, 320 ft. (100 m) deep in places, and mixed with ash from the eruption. This hardened into what's known as "tuff" rock and, over the years, the wind and rain has carved it, shaping deep valleys and tall columns of rock known as fairy chimneys.

DID YOU KNOW?

In Turkish, the fairy chimneys are called *Peri Bacalari.*

THREE BEAUTIES

Try to spot three iconic fairy chimneys, the "Three Beauties" (right), which are named after the ancient Greek goddesses Aphrodite, Athena, and Hera. Aphrodite was the goddess of love and beauty, Athena was the goddess of wisdom, warfare, and handicraft, and Hera was the goddess of marriage, family, and women. Their "heads" are formed from harder rock, sheltering the soft rock underneath and creating these unusual shapes.

THE STATS

Location: Cappadocia, Turkey
Best time to visit: May & Oct.
Height reached on flights: Up to 6,000 ft. (1,800 m)
Flight time: Sunrise
Length of flight: 1 hour
Age limit: 6 years and over

ROCK SHAPES

The Göreme Open Air Museum (below) is a must-see once you're back down on the ground. It's a cluster of churches, chapels, and monasteries cut into the rock. Some date back to the 11th century. Devrent Valley is another destination worth trekking to. With a bit of imagination you'll see all kinds of shapes in the rocks, including a camel!

MORE FUN!

Visit the Derinkuyu Underground City (left), a fascinating cave system hidden beneath Cappadocia. There are seven levels of tunnels and small caves, reaching down to 280 ft. (85 m), which once housed more than 20,000 people plus their livestock! Constructed over 3,000 years ago, they were forgotten about until the 1960s, when a local man happened to be doing some work on his house – only to discover a tunnel leading to cave rooms under his land! To get a feel of what life might have been like down there, reserve a room in a cave hotel in the area (luckily they now have all the modern conveniences!).

Climb to the top of the world's tallest building...

...AT THE BURJ KHALIFA

Welcome to Dubai, home of many modern architectural wonders, including the tallest building in the world – the Burj Khalifa! Get a true bird's-eye view of the whole country by taking a super-fast elevator to the observation deck near the top. But this is more than a one-hit wonder. The entire building is buzzing with activity: hotels, restaurants, shops, fountains... you can fill a whole day with fun!

OUTDOOR OBSERVATION

When heading to the outside viewing areas, expect a bit of a breeze. You are thousands of feet in the air, after all! It's also around 59°F (15°C) cooler up here, but that can be a blessing in Dubai. Time your trip for dusk and you'll see the Sun set on the desert and the city light up like a thousand stars. Planet Earth has many natural wonders, but man-made experiences like this can be pretty jaw-dropping too.

AT THE TOP

There are three observation decks inside the Burj, on the 124th, 125th, and 148th floors. On either of the two lower decks, use a telescope to take a close-up look at Dubai (far, far) below, before heading outside for a breath of high-altitude air. If you can stomach going even higher, level 148 really has the wow factor. Set 1,821 ft. (555 m) up, this is the highest observation deck in the world. You will feel like you are on top of the world!

DID YOU KNOW?

There are a whopping 2,909 stairs in the Burj Khalifa – what a workout that would be if you decided not to take the elevator!

THE STATS

Location: Dubai, United Arab Emirates
Best time to visit: Nov.–Feb.
Height: 2,717 ft. (828 m)
Floors: 163 in total
Cost: US$1.5 billion
Named after: Sheikh Khalifa, who helped fund construction of the building

ELEVATOR RIDE

The elevators in the Burj Khalifa are an experience in themselves. Traveling at a speed of 33 ft. (10 m) per second, they can reach the first observation deck in about a minute and carry up to 14 people at once. Inside, you'll be treated to a video giving a brief history of the Burj Khalifa.

WORLD RECORDS

The Burj Khalifa is not just the world's tallest building, it's a multiple record breaker. With 163 floors in total, it has the highest number of stories of any building. It also boasts the highest occupied floor, with apartments on level 154 and the tallest service elevator in the world.

MORE FUN!

Surrounding the Burj is the sprawling Dubai Mall. One of its main attractions is a huge aquarium. Outside the mall you'll find the spectacular Dubai Fountain (above). Tall jets of water shoot in time to music and lights in an impressive show that takes place every night of the year.

Experience a kaleidoscope of color...

...AT

HOLI FESTIVAL

On the day after the last full Moon of winter, the air across India becomes colored with bright neon powders as Holi is celebrated. Join this Hindu festival and usher out the cold winter, welcoming instead the arrival of spring and the hope of new life, new starts – and a lot of fun!

THE STATS

Location: Delhi and across India, and Hindu communities worldwide
Date: Usually in March, depending on the full Moon
Frequency: Annual
Type: Religious and cultural festival
Also known as: Festival of Color

ORIGINAL ORIGINS

This ancient festival has its origins in Hindu mythology. Holi comes from the name Holika, the sister of a demon king called Hiranyakashyap – the legend of whom reminds people that by staying on a righteous path they will be protected from harm. There are other stories associated with the festival too, and traditions vary across the country. Whichever legend is believed, they all essentially celebrate the triumph of good over evil.

PARTY TIME

Celebrations start the evening before, with the lighting of bonfires and the hope that any evil will be destroyed in the flames. The next morning, the Holi festival begins. Vibrant *gulal* powders are thrown into the air, along with water, making the colors stick to everything and everyone, including you! Singing, dancing, and drumming enhance the merriment. You'll be a rainbow of color during the day, but in the evening, it's time to get clean, dress up, and visit friends and family to celebrate and share food. One snack to try is *gujiya* (below) – a sweet pastry stuffed with a tasty filling of nuts and dried fruit.

WATER FIGHTS

In the ten days leading up to the festival, have fun practising your aim with water guns (called *pinchkaris*) and water balloons by hiding in the streets and trying to soak passers-by. Don't worry, you won't get into trouble – there's a popular saying, "*Bura na mano, Holi hai!*" that means "Do not mind, it's Holi!"

DID YOU KNOW?

Holi has been celebrated in India and neighboring countries for centuries. Other countries that have Holi celebrations include Nepal, Fiji, Mauritius, Guyana, Bhutan, Trinidad and Tobago, Pakistan, and the Philippines.

THE MEANING OF COLOR

Gulal powder (below) is often made from dyes found in nature. Red is the most used color – it symbolizes love and fertility and is used in weddings. Yellow, made from turmeric, represents health and happiness, while green stands for nature and new life. Pink, orange, purple, and blue dyes are also used – but never black or white.

MORE FUN!

India is a country of celebratory festivals. If you want to experience another, visit in October or November for Diwali – the Festival of Lights. Diwali is celebrated by lighting firecrackers, lanterns, and diya lamps and spending time with family and friends, eating and dancing. It lasts for five days and marks the start of the Hindu New Year.

Spot a tiger on safari...

...IN RANTHAMBORE NATIONAL PARK

Visiting Ranthambore National Park in Rajasthan, India, offers a rare privilege... to be able to spot a Bengal tiger in its natural habitat. These majestic beasts are the largest of all the big cats, and one of the most beautiful, with their distinctive stripy orange coat. But, as nocturnal hunters, they are hard to spot in the day, so part of the magic of this tiger safari is trying to find them, hidden in their forest habitat.

HIDE AND SEEK

You'll need to connect with a guide to help you find a tiger. These skilled locals know how to identify pugmarks (paw prints) on the forest floor, and they'll also listen for the alarm call of birds and monkeys as they alert other animals to the possible presence of the elusive big cat. Patience is required, but the wait builds suspense. When you finally spot a flash of orange, or the flick of a tail, it's an exhilarating moment!

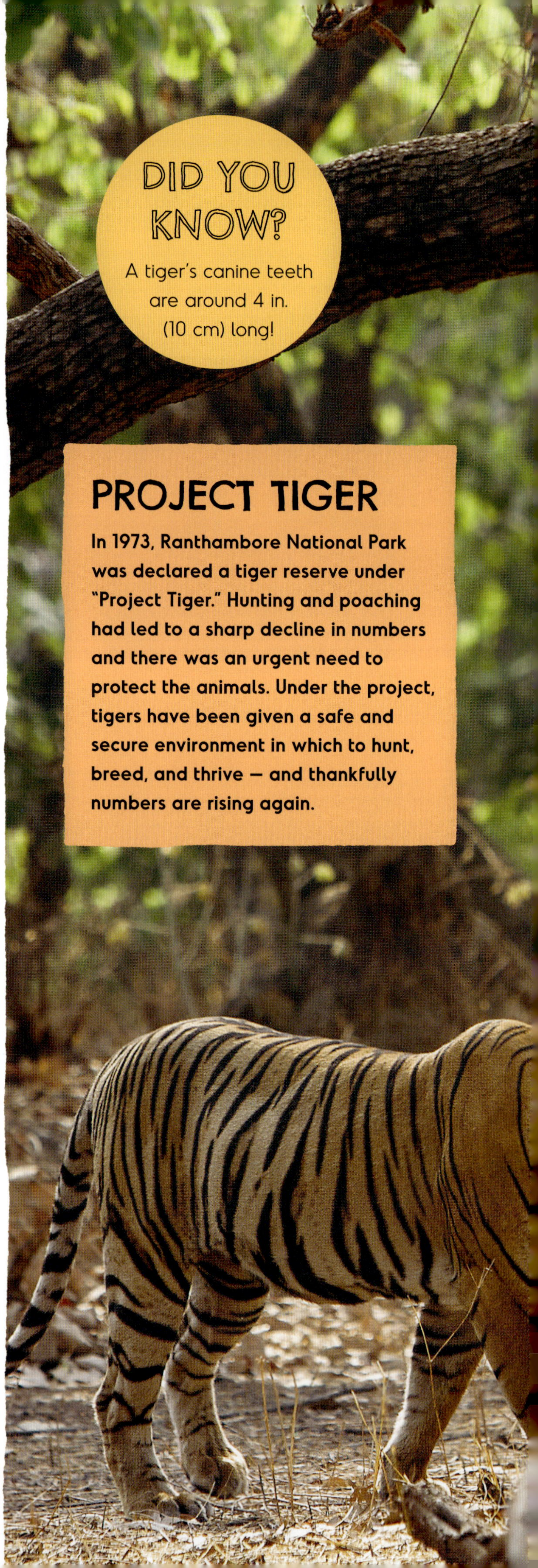

DID YOU KNOW?

A tiger's canine teeth are around 4 in. (10 cm) long!

PROJECT TIGER

In 1973, Ranthambore National Park was declared a tiger reserve under "Project Tiger." Hunting and poaching had led to a sharp decline in numbers and there was an urgent need to protect the animals. Under the project, tigers have been given a safe and secure environment in which to hunt, breed, and thrive — and thankfully numbers are rising again.

THE STATS

Location: Ranthambore National Park, Rajasthan, India
Best time to visit: Oct.–Jun.
Established: 1955
Size: 515 sq. miles (1,334 sq km)
Tiger safari length: 3.5 hours
Animals to spot: Bengal tigers, leopards, sloth bears, striped hyenas

TIGER TRAITS

The period from March to April, when the temperature is rising, is one of the best times for tiger-watching. Big cats feel the heat just like humans. They will emerge from the shade and seclusion of the bushes in search of water – meaning you have a better chance of spotting one. At other times of the year, tigers are mostly active at night, and use the cover of darkness to hunt buffalo, deer, and other large mammals.

SAFARI TIPS

- **Wear neutral clothing when out on safari (tigers are sensitive to reds and pinks), and be sure to cover up with sunscreen and a hat.**
- **When you see a tiger, keep still and don't make any sudden movements, as you might startle it.**
- **Don't feed the animals!**
- **Only ever take photos and memories away with you – leave the tigers' habitat untouched.**

MORE FUN!

Of course, a trip to India wouldn't be complete without a visit to the Taj Mahal, around 150 miles (250 km) to the northeast of Ranthambore National Park. This stunning monument was built by Emperor Shah Jahan for his wife, Mumtaz, as a symbol of his eternal love.

Climb the world's highest mountain...

...AT MOUNT EVEREST

Climbing Mount Everest is often seen as the ultimate physical challenge, which is why people have to be at least 16 years old to be allowed to climb it. However, that doesn't mean you can't enjoy part of the experience. A trek to the Everest Base Camp is something that can be attempted by all ages — with the right preparation — and is an incredible adventure in its own right.

ANCIENT FORMATION

Mount Everest, called Sagarmatha in Nepali, can be found on the border of Nepal and China. It began to form between 50 and 60 million years ago when two tectonic plates (the giant rock plates that make up the Earth's crust) collided, pushing the rock upward into a range of mountains. The pressure is still pushing them up today, so the summit actually rises about 0.25 in. (0.6 cm) each year. You'd better climb it soon, before it gets too tall!

THE STATS

Location: Nepal–China border, Himalayas
Best time to visit: Apr.–May
Height: 29,032 ft. (8,849 m)
South base camp height: 17,598 ft. (5,364 m)
North base camp height: 16,900 ft. (5,150 m)
Age limit: 16 years and over

BASE CAMP

Every year, mostly in spring and fall, a city of tents appears at the bottom of each of the main routes to Everest's summit. The northern base camp is next to the Rongbuk Glacier, and the southern base camp is next to the Khumbu Glacier. Both base camps are still at a very high altitude, but they are located just below the limit at which the air gets too thin for humans to survive long-term.

GETTING TO BASE CAMP

If you choose to visit the southern base camp, a week of trekking beforehand is required to reach it. It's hard work, but is mostly flat terrain, and you'll get to walk through national parkland, forests, and traditional villages (left) along the way. However, if you choose the northern base camp, you can actually drive the entire way there! At these base camps you'll find a few basic home comforts – showers, dining areas, and even internet access – but the main purpose of staying here is to allow climbers to adjust to the altitude and prepare for their trek to the summit.

PREPARING FOR THE CLIMB

At base camp, you'll meet many people who hope to make it to the very top. But they can't just march straight up. First they'll set up mini-camps along the route, containing life-saving supplies in case of avalanches, frostbite, or other weather extremes. Each climber will carry an oxygen tank as there is less air to breathe the higher you go. Altitude sickness can affect anyone, so be sure to take it easy even at base camp and enjoy some slow time just admiring the world's tallest mountains.

TAKE YOUR TIME

For those old enough to climb the mountain, the whole journey will take about two months. This involves traveling to base camp, acclimatizing for a few days, then spending around 5–6 weeks climbing up and then down between camps along the way, before returning to base camp feeling accomplished with your awesome achievement!

MORE FUN

While you're in the area, check out the bustling city of Kathmandu, Nepal's capital. In this largely Hindu country, cows are considered sacred and they're generally allowed to roam where they please. Once you've got used to seeing cows wandering the streets, raise your eyes upward and you'll see hundreds of colorful flags dancing in the wind. They are there to bless the countryside with prayers: blue for sky, white for air, green for water, yellow for earth, and red for fire.

Meet an ancient army...

...AND PLAY WITH PANDAS IN CHINA

China is one huge country – the third-largest in the world by area and the second-largest by population. It offers something for everyone, but the variety is so vast it can be hard to know where to start. From the ultramodern to the almost untouched, it's a country that honors its deep-rooted values, traditions, and myths, while also pushing boundaries in the modern world. Here are some highlights you should try and check off to get a feel for what it's like to live in China today.

SHANGHAI'S SKYLINE

The city of Shanghai is known for its glitz and glamor and quest to embrace everything modern. Walk along the Bund – the waterfront pathway beside the Huangpu River – and imagine Shanghai's transformation from an ancient port receiving deliveries of rice to the home of successful businesses that operate from the huge skyscrapers you see today. The Shanghai Tower (above center) twists its way like a dragon's tail to 2,073 ft. (632 m) and is the tallest building in the country.

TERRACOTTA ARMY

Come face-to-face with an army of soldiers in Xi'an – one of the world's most amazing archaeological finds. In 1974, a group of farmers discovered over 8,000 life-size statues made of clay. Each figure is unique – look closely and you'll see a range of clothing, hairstyles, and facial expressions. They were built for Qin Shi Huang, the first emperor of China, to protect him in the afterlife following his death in 210 BCE. It's thought that approximately 700,000 workers took 40 years to build all the soldier statues, as well as 130 chariots and 520 horses.

ROBOT RESTAURANT

Chinese cuisine is popular worldwide, but this restaurant is unlike anything else... Foodom in Guangzhou is staffed by 46 robots, and zero humans! Robots handle every step of the process in the restaurant, from receiving deliveries of ingredients to storing produce and then selecting the items required to make each dish. There are 32 chef robots and you can watch them hard at work behind glass windows. Meals are then whizzed from kitchen to table along a pathway of rails on the ceiling. The robots even clean up afterward, too!

THE STATS

Location: China

Best time to visit: Apr.–June, before it gets too hot in the south

Population: 1.4 billion

Capital: Beijing

Biggest city: Shanghai, with a population of 24 million

PLAY WITH THE PANDAS

Giant pandas are native to China, and are a national symbol. Their cute appearance and playful character makes them a favorite with tourists and locals alike, and their conservation is a top priority. Visit the Chengdu Research Base and see baby pandas hand-reared by keepers – they are tiny, pink, and hairless when they are first born. Or head into the mountains to find the Dujiangyan Panda Reserve, where pandas are trained to live in the wild so they can be released into their natural habitat. Wherever you are lucky enough to see them, these cuddly giants are sure to be munching on bamboo shoots and leaves.

Travel back to the age of emperors...

...IN

BEIJING

Welcome to the capital of China – one of the best cities in which to experience traditional Chinese culture and customs. Beijing is also the starting point for most visits to the Great Wall of China, an epic feat of engineering as well as a pathway through some stunning scenery. Enjoy life as the Chinese emperors of the past would once have done, by relaxing in picturesque gardens during the day and watching acrobats perform at night.

GREAT WALL

More than just a wall, this enormous structure is a collection of beacon towers, barriers, barracks, garrisons, and fortresses, which together formed a huge defense system to protect the northern part of the Chinese empire from enemy attack. At 13,170 miles (21,195 km) long, it's still the longest human-made structure ever built. Building began over 2,000 years ago, but most of the surviving sections were constructed during the Ming Dynasty, from the 14th to the 17th centuries.

BEST AREAS TO VISIT

The Great Wall of China doesn't go through Beijing itself but is north of the city. These three sections of the Great Wall can be easily reached via public transportation from your base in Beijing.

Juyongguan is a showcase for old temples and towers, as well as views.

Badaling is the best-preserved and most-visited section of the Great Wall.

Mutianyu has some serious scenery and fewer crowds.

FORBIDDEN CITY

China's largest and best-preserved collection of ancient buildings can be found in the Forbidden City in Beijing. Built in the early 15th century, it was reserved solely for use by the imperial family until 1912, hence the imposing name. Luckily, non-royals can enter the palace complex today! See how many dragons you can count in the Hall of Supreme Harmony where the throne sits (it may take you a while – there are over 10,000!).

DID YOU KNOW?

Dragons are an important symbol of Chinese culture, representing power, luck, strength, dignity, and wisdom.

SUMMER PALACE

A fun day trip out of the city is to the exquisite Summer Palace. Don't be confused by the name – this is actually a garden, begun in the 12th century, and was a favorite spot of the emperors when they wanted to escape the heat of the city in the summer months. You can ride a dragon-shaped boat on the enormous Kunming Lake (left) before exploring the temples, residences, pavilions, and bridges. Longevity Hill, with its three-leveled tower at the top, leads down to the Long Corridor, an ornamental walkway covered in thousands of artworks showing scenes from Chinese myths and folktales.

THEATER OF ACROBATS

The Flying Acrobat Show at the Chaoyang Theatre will provide you with some thrilling evening entertainment. The Chinese are well known for their acrobatic skills – they've practiced this art form for more than 2,000 years and it was a favorite show in the emperor's court, as well as loved by the masses. You'll get to see some amazing feats of acrobatics; contortion acts, lion dances, fire stunts, and artistic cycling... have you ever seen 10 people riding one bike before?

THE STATS

Location: Beijing, China
Best time to visit: Apr.–May & Sept.–Oct.
Population: 21.5 million
China ruled by emperors: 221 BCE–1911 CE

Watch snow monkeys enjoy a hot bath...

...AT JIGOKUDANI MONKEY PARK

In the forests of Jigokudani in Yamanouchi, central Japan, snow falls silently into a valley where steam rises from hot spring baths called onsens. It's a picture-postcard scene, and it's not just people who enjoy it – the local "snow monkeys" come here to relax as well. Japanese macaques are the most northern-living primates and, while their thick fur helps keep them warm when it's cold, they like nothing better than a long, hot soak in the bath!

ANIMAL INTERACTIONS

Be sure to dress for the cold when visiting in winter as temperatures can dip as low as 14°F (-10°C). The monkeys are fed daily by the park wardens – check the times when you get there if you want to watch. But it's important to remember they are still wild animals, so don't offer them any other food. The monkeys are very used to their human visitors and go about their business unfussed by the attention. They live in troops of up to 100 individuals, including babies, each led by an alpha male. You can check out portraits of the alpha males, dating back dozens of years, on display in the information center.

COPYCATS

Monkeys usually stay away from water, so how did this community of primates discover they enjoy bathing? The explanation could be as simple as the saying "Monkey see, monkey do...," in that they saw humans relaxing in the restorative hot springs and thought they'd try it too. Or maybe a monkey fell into the water one day, realized it was lovely and warm, and so didn't want to get out!

DID YOU KNOW?

Jigokudani Valley means "hell valley," which is a common name for areas with volcanic activity and steam rising from the ground.

THE STATS

Location: Jigokudani Yaen-Koen Snow Monkey Park, Japan
Best time to visit: Dec.–Mar., when snow falls
Established: 1964
Height: 2,789 ft. (850 m)
Open: Every day of the year

HOT WATER

Onsens are hot! The volcanically heated water often reaches temperatures of 102–107.5°F (39–42°C). This is because Japan sits on the boundary of four tectonic plates where hot magma from deep inside the planet rises to near the surface and heats water. This is also why Japan is a country of volcanoes and earthquakes – there's a lot of movement where the plates meet each other. But all this geological activity means that rich minerals are released from inside the planet, making the spring water good for your health.

ONSEN RYOKAN

Alongside many of Japan's natural hot springs you'll often find a traditional inn – an *onsen ryokan*. Reserve an overnight stay and you'll experience age-old Japanese customs and impeccable hospitality. Soak in the warm mineral-rich waters for as long as you wish, before drying off and sitting yourself down on tatami mat flooring for a delicious traditional meal eaten from a low table. Sleep on a futon mattress to recharge your batteries, then in the morning head back to the onsen for another dip before you check out. Bliss!

MORE FUN!

Just 20 minutes from Jigokudani Monkey Park is the country's largest ski resort. Shiga Kogen (below) has 18 ski areas and over 90 trails to navigate. You can ski, snowboard, or sled here from mid-November all the way through to late April. You'll find Japan's highest chair-lifted ski run here, too, taking you up to 7,569 ft. (2,307 m) above sea level!

Prepare to be amazed by the future...

...IN PRESENT-DAY TOKYO

When you think of a city of the future, you're probably imagining everything Tokyo already has to offer. It's the most populated city in the world, full of tall towers and bright lights. There's a solution to help every part of daily life run smoothly in this mega-metropolis... Tokyo is at the forefront of innovation, often leading the world in hi-tech inventions, electronics, and robotics.

SCIENCE AND SCI-FI

You can see a range of cool technology from past to present at the National Museum of Nature and Science. Or if you like things a little more sci-fi, head to the teamLab Planets art gallery (below). Here, you can experience a mix of virtual and augmented reality as you immerse yourself in art projected into the rooms. In the water area you can take off your shoes and wade barefoot, while interacting with digital fish projected onto the surface.

TECHNOLOGY EVERYWHERE

What does a futuristic city look like to you? Tokyo is already way ahead of the curve, with state-of-the-art details in almost everything here. Vending machines on street corners sell drinks and snacks as you would expect, but they also sell umbrellas, T-shirts in a can, and surprise gifts. Then there are the toilets — are you ready to experience seat-warming, massage, sprays, music, and even scent options? And you won't have to look too far to find a robot in a public place providing entertainment and information for passers-by.

THE STATS

Location: Tokyo, Japan
Best time to visit: Apr.–May & Sept.–Oct.
Population: 37 million
Vending machines in Tokyo: 4 million

DID YOU KNOW?

One of the world's most futuristic hotels is in Tokyo. The Henn-na Hotel is staffed by robots!

SUPER SHINKANSEN

Although they are now over 60 years old, Japan's *shinkansen*, or "bullet trains," are still futuristic icons, zipping about the country at speeds of up to 200 mph (320 km/h). But riding one is not just about speed, it's an experience. Your seat is designated, your travel is relaxing, and you're encouraged to bring a delicious bento box meal, called an *ekiben*, on board with you. And you definitely won't be late – everything runs like clockwork on these lines!

TALLEST TOWER

Another symbol of Tokyo's modernity is the Skytree, the tallest tower in Japan at 2,080 ft. (634 m) and a broadcasting hub. You'll get a fantastic 360-degree view from the observation deck 1,476 ft. (450 m) up (right). On a clear day you'll even get to see into the past... majestic Mount Fuji (top right), a symbol of the country, is an active volcano that was formed over 100,000 years ago.

MORE FUN!

Of course, old still meets new in some areas of the city. Head to Asakusa to see the traditional side of Tokyo. Visit the Sensoji Temple and pass through the Thunder Gate with its giant paper lantern (right). Then browse the traditional craft shops and snack on a *tennen taiyaki* (fish-shaped waffle filled with red bean paste).

Marvel at staggering Supertrees...

...AT THE FUTURISTIC GARDENS BY THE BAY

Discover art mixed with nature! This showcase of the natural world in Singapore is a collection of over 1.5 million plants from all over the globe. But what makes these gardens unique is how many of the plants grow on huge artificial structures called Supertrees. It's a magical experience to walk among vertical gardens, even more so at night when they are beautifully lit up!

SUPERTREE GROVE

There are 18 Supertrees to be found in Gardens by the Bay, ranging in height from 82–160 ft. (25–50 m). Head to Supertree Grove to come face-to-face with 12 of them. Here you will discover 162,900 plants representing over 200 species. After experiencing the plant life from the ground, head to the observatory on the tallest Supertree for city views and immersive digital displays.

DID YOU KNOW?

Environmental sustainability is high on the priority list at Gardens by the Bay – the park is entirely powered by solar panels positioned on the top of seven Supertrees.

SKYWALK

Take the skywalk path to see the plants growing from above. You'll spot many colorful blooms – including bromeliads and orchids – as well as tropical climbing plants. They thrive in Singapore's balmy climate, with year-round warm temperatures and high humidity.

THE STATS

Location: Singapore
Best time to visit: Dec.–Jun.
Size: 0.4 sq. miles (1.01 sq km)
No. of plants: Over 1.5 million
Height of the Supertrees: 82–160 ft. (25–50 m)

NIGHT LIFE

Every evening, the Supertrees "come alive," putting on a light and sound show known as "Garden Rhapsody." Lights dance across the trees to a musical soundtrack during the 15-minute performance.

A GARDEN CITY

Although the iconic Supertrees are the highlight of the gardens, don't forget to visit the other natural sanctuaries here. Kingfisher Wetlands consists of a freshwater lake with cascading water and streams to attract migrating birds. And inside the Flower Dome, officially the world's largest greenhouse, you'll find waterfalls, a cloud forest, rare flowers, and even a collection of poison dart frogs. Gardens by the Bay also hosts festivals, concerts, movie screenings, and sporting events, so there really is something for everyone!

CHILDREN'S GARDEN

One final area you won't want to miss is the Children's Garden. It's full of interactive delights – a swaying bridge, stepping stones, and rainforest tree houses featuring a maze of ropes, ladders, and slides. After a hot day of exploration, head straight to the water play area (above) and splash to your heart's content. Watch out for the motion-sensor fountains as you whizz past.

MORE FUN!

Singapore is both a country and a city, so it's very compact and easy to get around. After visiting Gardens by the Bay, head to a hawker center (food market) and try a traditional lunch, such as locally-caught chili crab. Once your food's gone down, head to Sentosa Island to spend some time on the beach, or enjoy a cable car ride and the Skyline Luge. After sunset, head to Singapore Zoo for a night safari to observe nocturnal animals going about their business!

Make a wish...

...AT THE LANTERN FESTIVAL

Incredible experiences don't have to be full of adrenaline and adventure. Sometimes the simplest of events can leave the biggest impression, and the Full Moon Lantern Festival in Hội An, Vietnam, will do just that. On the 14th day of each lunar month, when the Moon is at its fullest and brightest, the lights are switched off throughout this ancient town and instead the warm glow of thousands of lanterns illuminates the river and streets.

RIVER LIGHTS

As night falls, buy a flower lantern from a local, light the candle inside and make a wish as you release it down the Hoai River. These lanterns are made from colored paper – often red, as this symbolises love, happiness, and luck. Once you've set your lantern off, you can travel on a sampan boat alongside the floating lanterns to watch the progress of the twinkling candlelight.

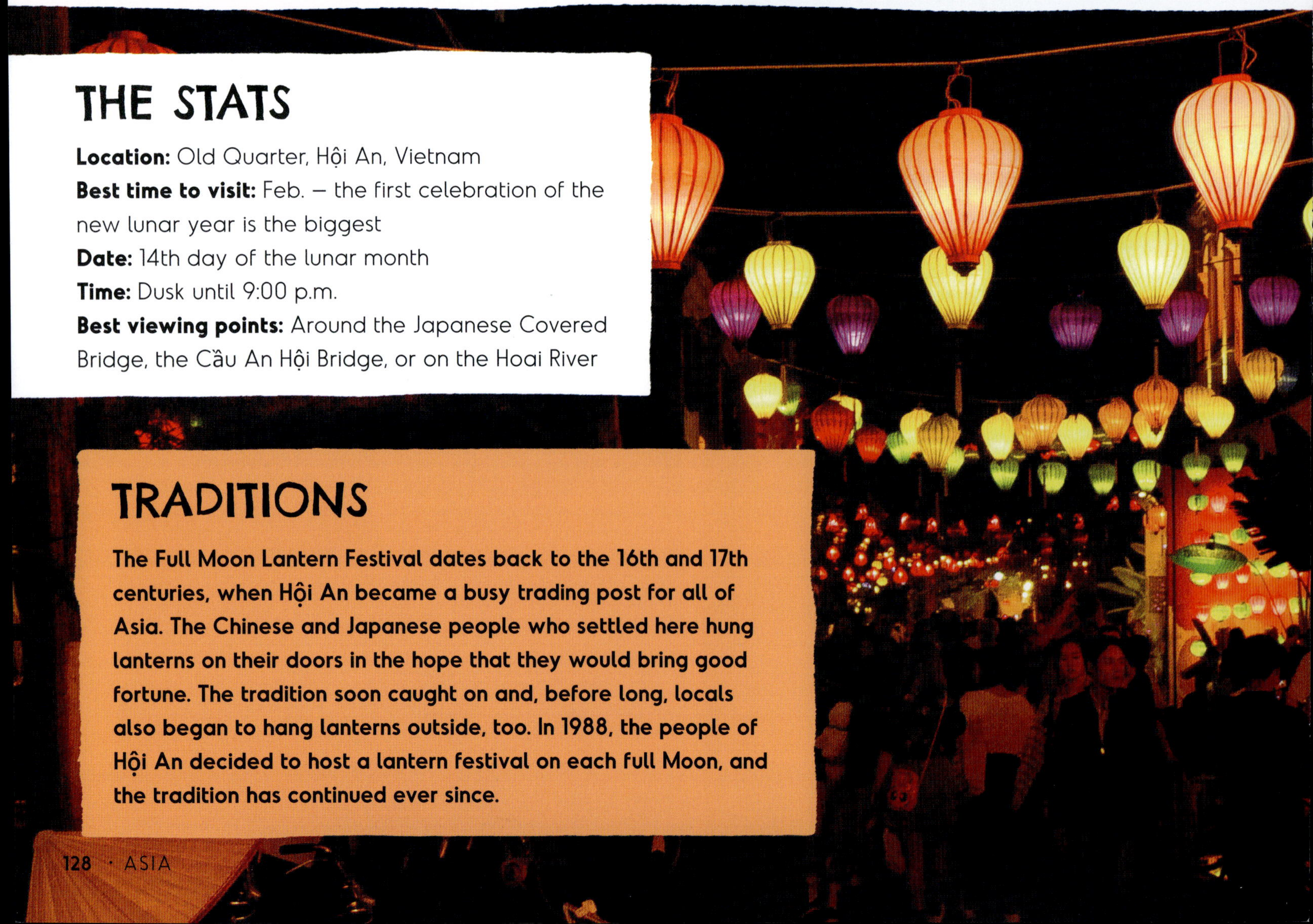

THE STATS

Location: Old Quarter, Hội An, Vietnam
Best time to visit: Feb. – the first celebration of the new lunar year is the biggest
Date: 14th day of the lunar month
Time: Dusk until 9:00 p.m.
Best viewing points: Around the Japanese Covered Bridge, the Cầu An Hội Bridge, or on the Hoai River

TRADITIONS

The Full Moon Lantern Festival dates back to the 16th and 17th centuries, when Hội An became a busy trading post for all of Asia. The Chinese and Japanese people who settled here hung lanterns on their doors in the hope that they would bring good fortune. The tradition soon caught on and, before long, locals also began to hang lanterns outside, too. In 1988, the people of Hội An decided to host a lantern festival on each full Moon, and the tradition has continued ever since.

STREET FOOD

Food is a must at any festival, and there are plenty of amazing delicacies for you to try here! *Cao lầu* is a pork noodle dish served with a unique local sauce, and *bánh mì* (below) is a baguette stuffed with a spicy, sweet, and salty filling that will tickle your tastebuds. *Com gà* is chicken rice, and *mì Quang* contains flat rice noodles in a broth. But if you have a sweet tooth, you may prefer to try a pretty mooncake, filled with a sweet red bean paste. Yum!

PARTY GAMES AND PERFORMANCE

Alongside the river, you will find plenty of entertainment, particularly in the form of the charming Bài Chòi folk art game. Similar to bingo, players sing songs to match images on cards. If the song is good, the player receives a flag, and the player with the most flags wins. This traditional performance game combines music, poetry, and acting, and is now on UNESCO's World Heritage List.

DID YOU KNOW?

The full Moon is also a time for locals to meditate, reflect on life, and worship their ancestors, so many will visit shrines to remember their loved ones.

OTHER FESTIVALS

Every Vietnamese child loves celebrating Tet Trung Thu Festival, also known as the Mid-Autumn Festival, or Children's Festival. The celebrations are partly to mark the end of the harvest, when the hard work is over. But it's also believed to be linked to the story of a man called Cuoi, who held on to a magical banyan tree as it floated up to the Moon. It's said that if you look closely at the Moon, you'll see the shadow of a man sitting under a tree. So each year, children parade lanterns in the streets to help Cuoi find his way home again. Maybe even more importantly, it's a chance to eat as many mooncakes as your belly can hold!

Next, it's...

...OUTSTANDING OCEANIA

The huge landmass of Australia dominates this region, and its famous animals will be high on your experience bucket list. Travel along the east coast to feed jumping crocodiles, swim with the "great eight" of the Great Barrier Reef, search for whales off the coast of New South Wales, and meet the many marsupials of Kangaroo Island. You'll get even more of a workout in New Zealand as you explore caves, bathe in mud pools, and take on some action-packed adventure sports like bungee-jumping. The South Pacific Ocean will help you relax at the end of an epic trip by offering unique chances to swim with manta rays and release rescued baby turtles back into the sea.

PLACES TO VISIT:

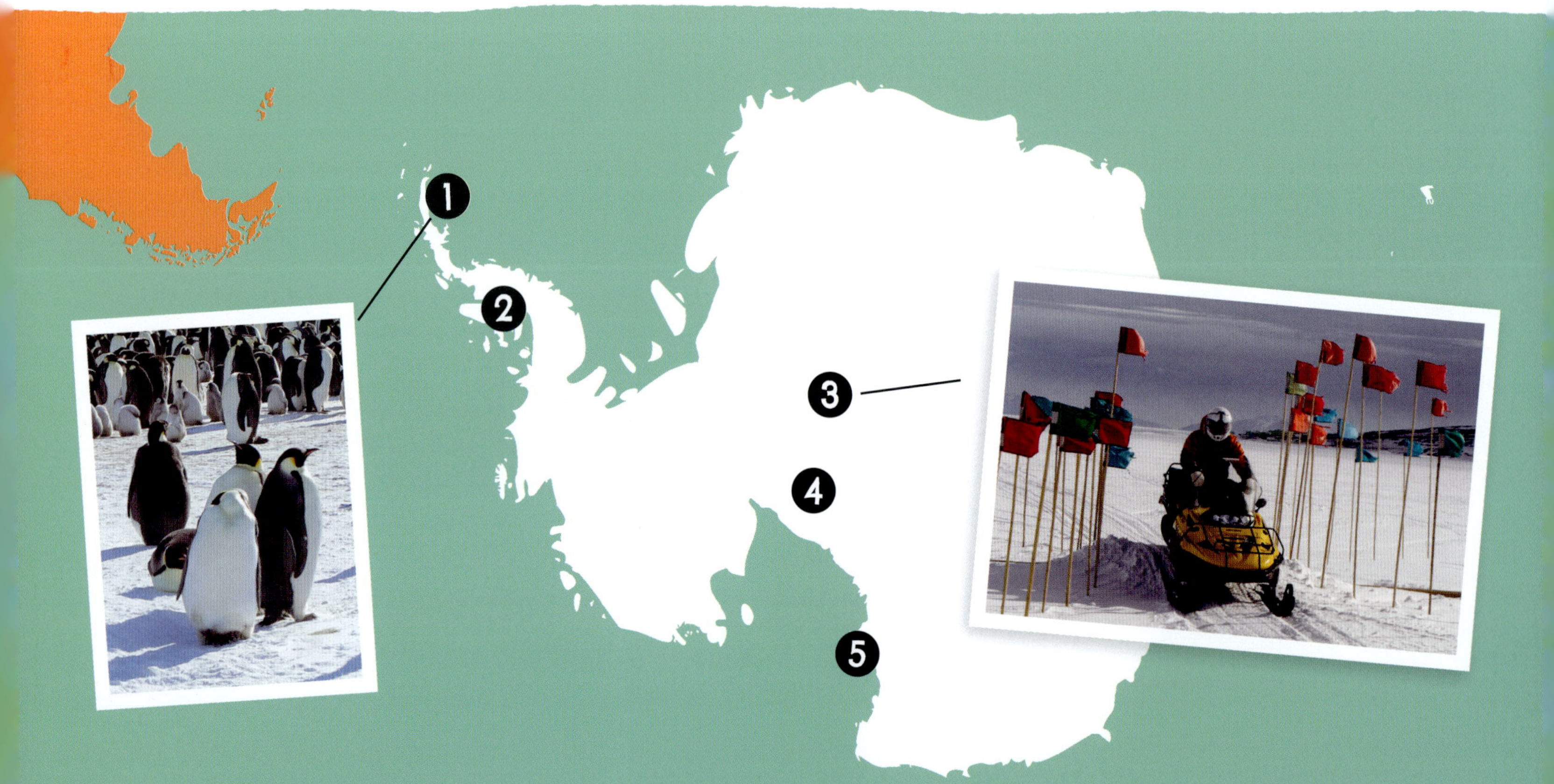

DID YOU KNOW?

There are twice as many kangaroos as people in Australia.

And finally, it's...

...ASTOUNDING

ANTARCTICA

After everything, don't forget about our special seventh continent. It's the coldest and most extreme of them all, but offers some of the most incredible experiences, from meeting penguins to riding on snowmobiles. Half the fun will be just battling the elements to get here. And with just a handful of tourists for company, you'll be making some truly unique memories.

PLACES TO VISIT:

1. Snow Hill Island, p.154
2. Cuverville Island, p.154
3. Amundsen-Scott South Pole Station, p. 157
4. Transantarctic Mountains, p. 157
5. Shackleton's Hut, p. 157

Feed jumping crocodiles...

...ON THE ADELAIDE RIVER

Australia is famous for its unique animals. You've probably heard about the more dangerous ones, as well as those that are cute and cuddly! So, what better way to start your Aussie adventure than by coming face-to-face with one of the scariest beasts alive – the saltwater crocodile. Cruise down the Adelaide River in the far north of the country and test your luck on a lunch date with a jumping croc... hopefully you'll survive to tell the tale!

CROC CRUISE

Saltwater crocodiles are fearsome, dangerous predators, and were hunted to near extinction on the Adelaide River by the 1970s. However, as numbers dwindled, locals soon realized the crocs actually needed protecting, not hunting. Today, you can take a boat tour down the river with a professional guide and watch them feed crocodiles from the deck. They'll dangle meat from a stick several feet above the water for the crocs to jump out and grab. Be sure to have your camera at the ready – it sure is a sight to behold!

STRONG SALTIES

These crocs are the largest reptiles on Earth, and some of the best hunters around – they sit at the top of the food chain for good reason. The powerful beasts like to feed on everything from small fish to large cattle... and won't hesitate to attack humans too, if provoked. Much of their success is down to their patience – they will happily sit and wait in the muddy waters, camouflaged out of sight, for hours. But when unsuspecting prey comes close, it only takes a second for them to spring into action and attack. They have some of the strongest jaws in the animal kingdom and, once they bite, they won't let go.

THE STATS

Location: Adelaide River, Northern Territory, Australia
Best time to visit: Apr.–Sept., during the dry season
Tour length: 1 hour
Age limit: Suitable for all ages
Croc size: Up to 20 ft. (6 m)

HOW DO THEY JUMP?

Crocodiles have short legs, so you might wonder how they create enough force to leap out of the water. The trick is in the tail – this mighty muscle can be up to 7 ft. (2 m) long and is very, very strong. When ready to attack (or grab the dangling meat), all the croc needs to do is wave its tail from side to side and the energy created is enough to propel it up and out of the water. Some can even launch their entire body high into the sky!

MORE FUN!

While you're in the swing of croc-watching in Australia, you should try and visit another top crocodile habitat. Just south of the Adelaide River, you'll find the Kakadu National Park, where you can see crocodiles sunning themselves before they slip into the water to cool off. More than half of the park is Aboriginal land, where Indigenous people have lived for around 65,000 years. Rock art from 20,000 years ago can still be seen in the area – some of it, of course, depicts crocodiles!

Dive into a watery wonderland...

...AT THE GREAT BARRIER REEF

Slip on your flippers to explore the Great Barrier Reef – nature's most amazing waterpark. Often described as one of the Seven Natural Wonders of the World, this vast reef system is made up of over 2,500 coral reefs and 900 islands off the northeast coast of Australia. A wildlife hot spot, it's home to some of the most fascinating creatures on the planet – and there are endless ways you can explore it all.

GREAT SIZE

The Great Barrier Reef is so huge it can even be seen from space. It covers an area of 139,000 sq. miles (344,400 sq km) – that's about the same size as Japan. More than three-quarters of the world's coral species are found here. Unfortunately, global warming is damaging the reef and causing some areas to lose their colors and turn white – known as coral bleaching. If you're lucky enough to catch a helicopter flight, look out for Heart Reef, a natural reef 56 ft. (17 m) wide in the shape of a heart.

GREAT FUN

How would you like to explore the reef? You could try something different every day for a week! You can snorkel in the shallow reef waters or scuba dive out into deeper depths. You could then paddleboard or canoe on the surface for a bit before flying high in a seaplane for a bird's-eye view. You can even camp overnight on one of the reef's many deserted islands.

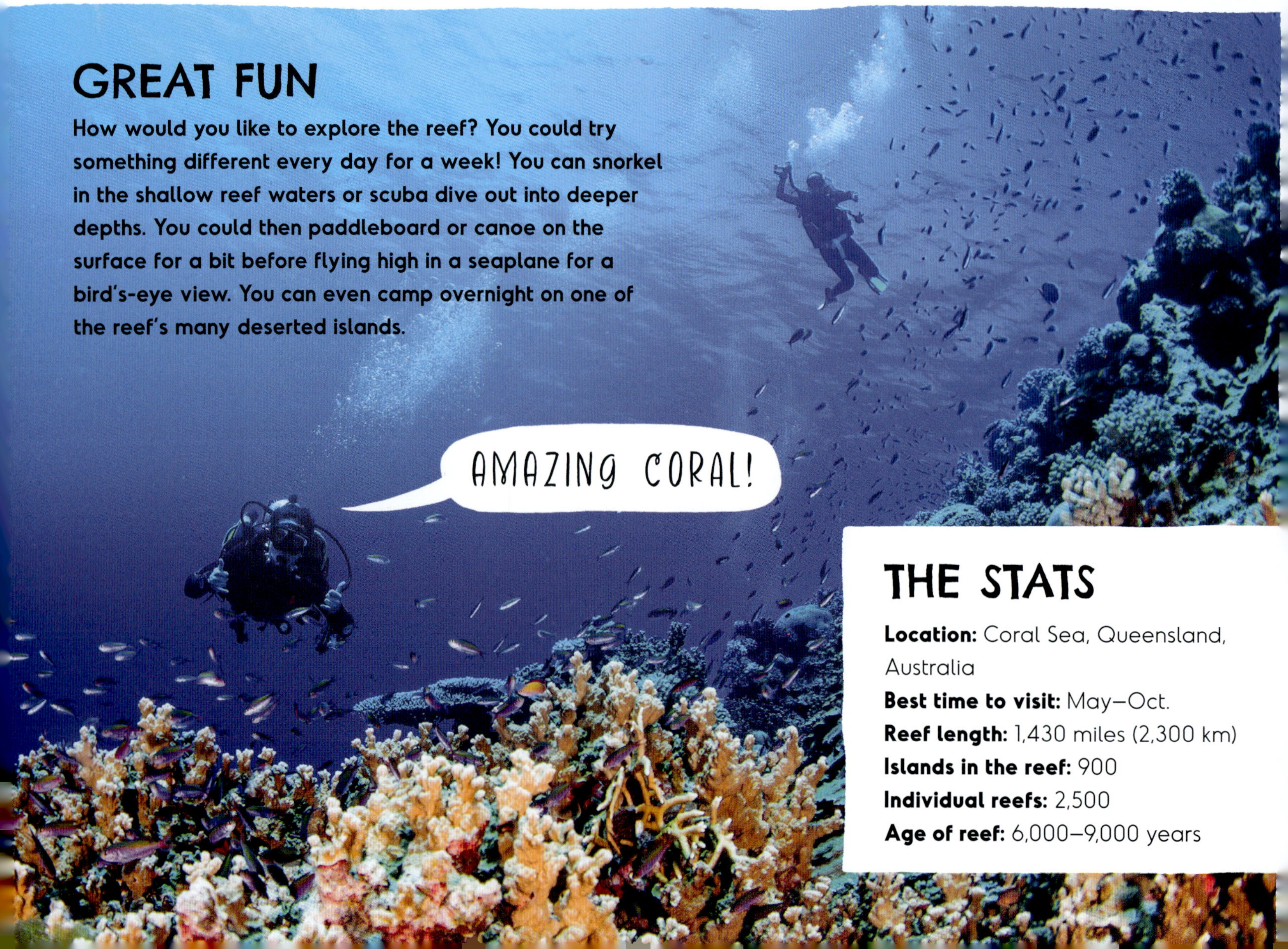

THE STATS

Location: Coral Sea, Queensland, Australia
Best time to visit: May–Oct.
Reef length: 1,430 miles (2,300 km)
Islands in the reef: 900
Individual reefs: 2,500
Age of reef: 6,000–9,000 years

GREAT EIGHT

You may have heard of the "Big Five" animals that everyone wants to spot when on safari in Africa... well, the Great Barrier Reef has the "Great Eight" for visitors to check off their list as well.

Shown above, the eight are:

1. Whale, 2. Sea turtle, 3. Shark 4. Clownfish, 5. Giant clam, 6. Manta ray, 7. Potato cod, 8. Humphead wrasse.

These are the species everyone wants to see. How quickly do you think you could spot them all?

GREAT BUILDERS

The reef system is made of corals, which come in many shapes and colors. Corals may look like plants, but if you get close up, you'll see they're actually communities of tiny animals known as polyps, which are similar to sea anemones. Some corals are soft, but hard corals take the nutrient calcium from seawater to make outer skeletons for themselves. It is these skeletons that join together to form huge reef structures.

GREAT (AND SMALL) FISH

To complete your reef experience, see if you can spot both its biggest and smallest inhabitants. Whale sharks (above) grow up to 40 ft. (12 m) long, making them the world's biggest fish. The world's second-smallest fish, the tiny stout infantfish, is also found here, but is a bit harder to spot at just 0.28 in. (7 mm) long.

Surf the Superbank...

...OFF THE GOLD COAST

Everything glitters in the aptly named Gold Coast in southeast Queensland. Its endless sandy beaches are constantly drenched in sunshine, and the whole city has become a world-class tourist destination. But one of its biggest draws are the waves that crash onto its beaches. Famous surfers come here to hit the beach and soak up the famous lifestyle, and the waves rarely let them down. In fact, the "Superbank," a human-made surf break, allows surfers to ride a single wave for an epic 1.2 miles (2 km)!

THE STATS

Location: Queensland, Australia
Best time to visit: May–Oct.
Superbank wave length: 1.2 miles (2 km)
Surf schools in Australia: 107
Other iconic surf spots: Manly Beach, Bells Beach, Margaret River

SURF SPOTS

The Gold Coast has many awesome beaches, but there are three in particular that draw surfers in their thousands to ride their waves, so be sure to visit one of these. Burleigh Heads, Snapper Rocks, and Kirra Beach have great conditions for long rides, and the crystal-clear water is very enticing on a hot day. Time it right and you might even catch an international surf competition during your visit!

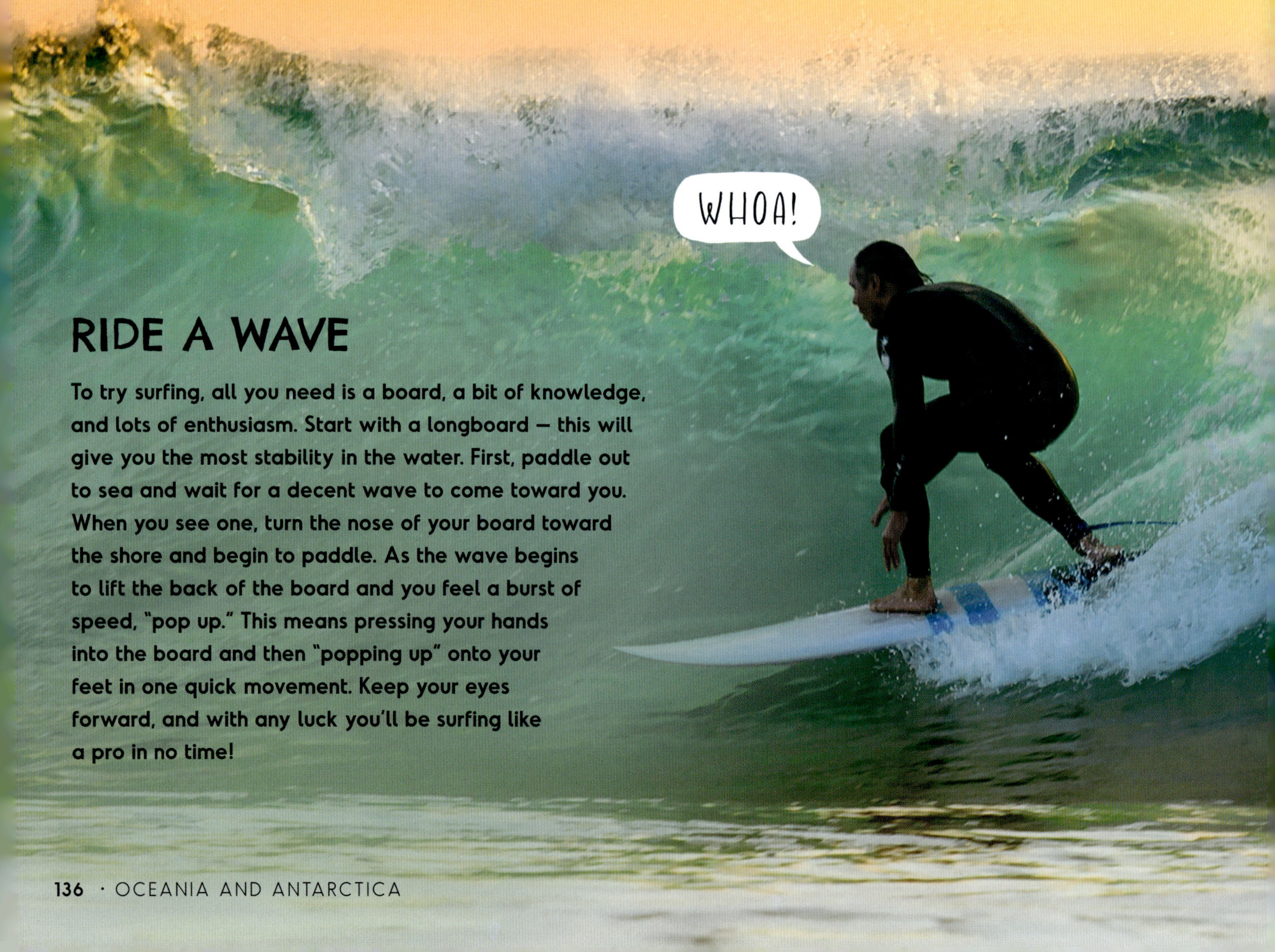

RIDE A WAVE

To try surfing, all you need is a board, a bit of knowledge, and lots of enthusiasm. Start with a longboard – this will give you the most stability in the water. First, paddle out to sea and wait for a decent wave to come toward you. When you see one, turn the nose of your board toward the shore and begin to paddle. As the wave begins to lift the back of the board and you feel a burst of speed, "pop up." This means pressing your hands into the board and then "popping up" onto your feet in one quick movement. Keep your eyes forward, and with any luck you'll be surfing like a pro in no time!

BUILDING A SANDBANK

Between the years 2000 and 2002, over 11 million sq. ft. (1 million sq m) of sand was dredged up and moved to the northern side of the Tweed River. The aim was to maintain a safe entrance to the river for boats, but it's become famous for the impact it's had on the surf!

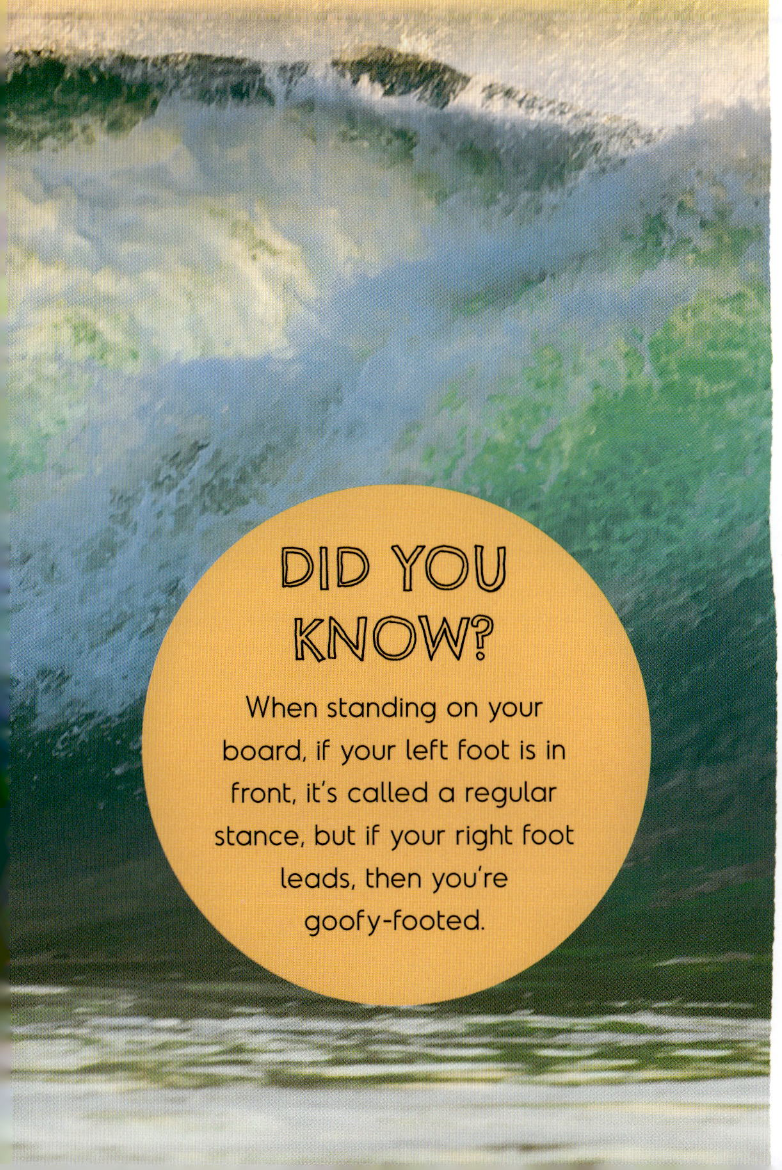

DID YOU KNOW?

When standing on your board, if your left foot is in front, it's called a regular stance, but if your right foot leads, then you're goofy-footed.

THE START OF SURFING

Surfing, and the easy, laid-back attitude to life associated with the sport, is all part of the Australian identity. The sport didn't originate here though. In 1915, a Hawaiian named Duke Kahanamoku brought his board to Sydney and showed the local people how to use it. And before that, the Polynesians used planks of wood to ride the waves.

NIPPERS

The sea is unpredictable, and even experienced surfers and swimmers can get pulled out of their depth quickly by rip currents. This is why many beaches are patroled by lifeguards, so be sure to stick to their rules. And if you're here for a while, why not join the local "Nippers" surf lifesaving group (above)? It's a fun way to enjoy the beach and waves while learning important water safety skills.

MORE FUN!

Not only famous for its beaches, the Gold Coast is also home to some of the country's best theme parks. Wet'n'Wild, Dreamworld, and Movie World can all be found in town. Or for a more natural adventure, you could climb up and slide down the largest sand dune system in the world at Moreton Bay (below).

Climb a bridge and catch a show...

...IN SYDNEY HARBOUR

THE STATS

Location: Sydney, Australia
Best time to visit: Sept.–Nov. for the weather; Dec.–Jan. for New Year's celebrations
Population: Over 5 million
Harbour Bridge opened: 1932
Opera House opened: 1973

You've arrived in Sydney, the multicultural city that is the sparkling jewel in the crown of the east coast of Australia. Everyone wants to be by the water here, and the city sprawls around a deep harbor. There's lots to see and do, so hop on board a cruise boat, ferry, tall ship, or kayak and take in the view!

OPERA HOUSE

The Harbour Bridge goes hand in hand with the iconic Sydney Opera House on every picture postcard of Sydney. This amazing building was designed and built between the 1950s and 1970s, and its white "sails" make it look like a ship – stand up close and you'll see they are covered in around 1 million tiles. Sydney Opera House is a performing arts hub, showcasing world-class acts of ballet, comedy, music, dance, and, of course, opera. But the outside can also put on a show. The sails double as a blank canvas, and lights are projected onto them to highlight important events or celebrations (right).

HARBOUR BRIDGE

Opened in 1932, this steel structure links the north and south sides of the harbor and is used by cars, trains, and pedestrians 24 hours a day. Affectionately nicknamed the "Coat Hanger," the bridge is not just a bridge – it's an experience! Rather than simply walk across it, why not try climbing the bridge (on an official tour, of course), cruising under it, or even gazing up at it from a swimming pool below? Once a year this structure attracts all eyes – a choreographed fireworks display is launched from it every New Year's Eve and the whole city stops to watch.

BEACHES AND BAYS

There aren't many destinations where you can be shopping or dining in a bustling city one moment and surfing or swimming on a pristine beach the next. This city offers calm harborside bays for you to take a relaxing dip and wave-filled fun at the beaches. Head down to the world-famous Bondi Beach early in the morning to see the sunrise and catch a wave before breakfast. Then trek along the headland to see the cliffs at the entrance to the harbor before turning into secluded Watson's Bay for a float in the still seawater followed by fish and chips as the Sun sets.

FUN ON THE FERRY

Sydney's green and yellow ferries are constantly zipping to and fro across the water. One of the most popular routes is from the hub at Circular Quay out to the beachside suburb of Manly on the north side of the harbor.

LUNA PARK

Can you spot a big smile under the Harbour Bridge on the north shore? It's the entrance to Luna Park – a vintage amusement park first opened in the 1930s. The iconic face is 30 ft. (9 m) wide and has had eight makeovers during its lifetime. Ride one of the roller coasters, enjoy the amusements, or hop on the Ferris wheel for another chance to view the harbor from up high.

Watch humpback whales...

...OFF THE COAST OF NEW SOUTH WALES

Could you find a needle in a haystack, or more to the point, a single humpback whale in the ocean? These marine mammals may be huge, but because they migrate distances of over 5,000 miles (8,000 km) twice a year, there's a huge amount of water to search. Luckily, these magnificent creatures are quite predictable – every year they appear off the east coast of Australia. So choose your boat, bone up on how to spot them, and head out on the whale-watching experience of your dreams.

CLEVER CLUES

Once you reach open water, keep your eyes peeled for signs. Whales travel in family groups called pods, so extra waves and splashes of white froth might be your first clue that they are close. Another sign to look for is spray. Whales breathe through blowholes on top of their heads – a big breath in allows them to stay underwater for up to 30 minutes. But when they come back to the surface for air, their exhale creates a spray of mist.

COMMUNICATION

As your boat draws near to where the whales might be, the driver will cut the engine. Can you hear anything now all is quiet? Some splashing in the water, maybe? Whales communicate by slapping the sea with their tails and fins. The more force they use, the more serious the message to be passed on – and the more likely you'll hear it! If you're really lucky, you might even pick out the call of a male. Their moans, howls, and cries can travel up to 20 miles (30 km) in water.

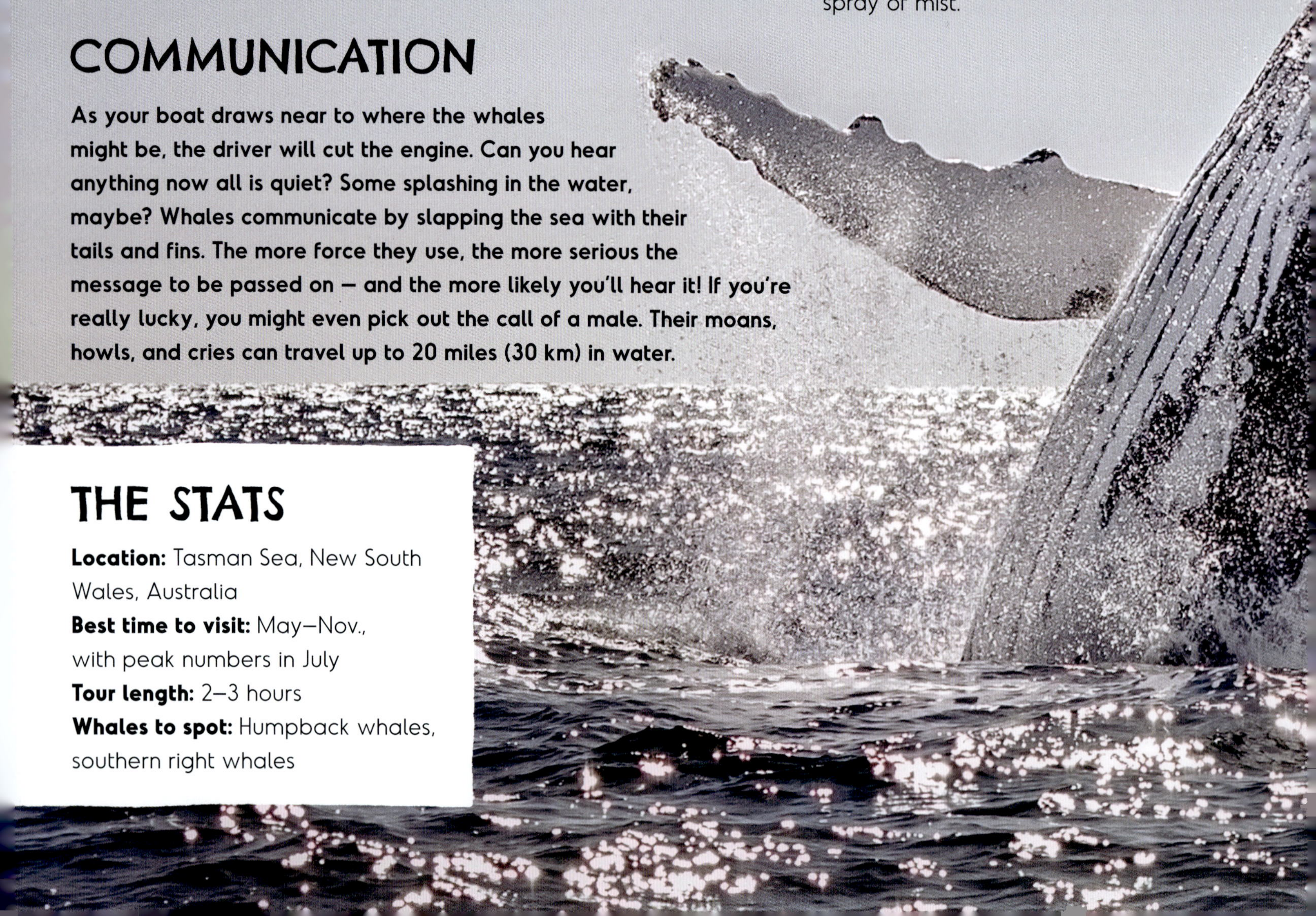

THE STATS

Location: Tasman Sea, New South Wales, Australia

Best time to visit: May–Nov., with peak numbers in July

Tour length: 2–3 hours

Whales to spot: Humpback whales, southern right whales

BREACHING

It'll happen suddenly, without any warning – a whale breaching out of the water – and you'll be there to witness the amazing sight! Breaching is when the whale leaps out of the water, allowing you to see it in all its glory. To lift such a huge weight requires serious power and speed. Humpbacks swim up to the surface and then use their tail to lift their body up and out of the sea, splashing back down on their back or side. It's a whole lot of fun for them... and for you!

DID YOU KNOW?

Humpbacks get their name from the hump on their back, in front of their dorsal fin.

HUNGRY HUMPBACKS

Humpback whales love to eat krill. These tiny shrimp are minuscule in comparison to the whales though, so they eat thousands of them with each gulp. Whales migrate from warm tropical waters, past Australia, and down to the cold Antarctic, where there's plenty of krill in the chilly waters for them to feed on. Catching krill is quite an art. The whales blow through their blowholes as they circle the krill, creating a net of bubbles around them. The krill bunch together and the whales gobbles them down in one huge gulp.

MORE FUN!

If you want some more high-adrenaline sightseeing on water, jump on a jet boat ride. Many of Sydney's sights are situated around the beautiful harbor, so viewing them from the water is ideal. Your jet boat will spin, slide, fish-tail, and power-break-stop, all to top tunes. Hold on tight and be prepared to get wet!

Meet the marsupials...

...BY HOPPING OVER TO KANGAROO ISLAND

You probably know that marsupials are cute, cuddly creatures that call Australia home. But how do you feel about meeting one in the wild? Could you tell a kangaroo from a koala, or a wallaby from a wombat? Head to Kangaroo Island, off the south coast of Australia, to experience some face-to-face encounters with Australia's best-loved wildlife.

READY TO SPOT A 'ROO?

Kangaroos' trademark bounce makes them the iconic Aussie animal. They come in various sizes – the biggest, the red kangaroo, can be up to 6 ft. (2 m) tall and clear 25 ft. (8 m) in one bounce. At the other end of the scale, the musky rat kangaroo is – rather unsurprisingly – about the same size as a rat! Kangaroo Island also has its own unique species, the Kangaroo Island kangaroo. Look out for groups of kangaroos, called mobs, courts, or troops, at sunrise or sunset.

MARSUPIAL MOMS

One of the unique features of marsupials is that they carry their young in a pouch. Only a few weeks after an egg is fertilized, a marsupial baby is born and crawls its way into the pouch on its mother's abdomen. Here it nurses on milk for several months until it is big enough to emerge from the pouch and start taking its first steps. But it'll often hop back into the pouch for a bit of comfort and rest!

DID YOU KNOW?

Kangaroos can't hop backward – their large feet and long tails won't let them move in that direction!

THE STATS

Location: Kangaroo Island, South Australia
Best time to visit: Sept.–Apr. for warmer weather
Size: 1,700 sq. miles (4,400 sq km)
Best places to see wildlife: Flinders Chase National Park, Kangaroo Island Wildlife Park, Hanson Bay Wildlife Sanctuary

GREAT GLIDER

Other marsupial residents of the island include wombats, wallabies, possums, and the "pygmy" or "feather-tailed" glider. The size of a small mouse, this is the world's smallest gliding mammal. It can make short flights through the air between trees using a bit of extra skin called a patagium that stretches between its forelegs and hind legs. Pygmy gliders can be found in forests all down Australia's east coast.

SPY A SLEEPY KOALA

Look up into a eucalyptus tree and with any luck you'll spot a cuddly koala curled up in the branches. The population on Kangaroo Island was introduced here, but they've settled in well and have survived challenges like bushfires and droughts. They eat only eucalyptus leaves, which iswhy they sleep up to 18 hours a day – they don't get much energy from their leaf-based diet.

MORE FUN!

Kangaroo island has some incredible physical features for you to discover. Visit the Remarkable Rocks (right), which really are, as their name would suggest, remarkable. Sitting in a prominent position on the headland, they've been shaped by the wind, sea, and rain for over 500 million years. Another landmark that can thank the elements for its shape is Admirals Arch, which has been formed by the power of the waves.

Let glowworms light your way...

...IN THE WAITOMO CAVES

Kia ora – *welcome to New Zealand! This is a country of mystery and intrigue, where nature sets the pace of life. For your first adventure here, head to the Waitomo Caves to explore a magical underground river system where a light show is put on by nature itself.*

RUAKURI CAVE

This cave, the largest in the region, is entered via an impressive spiral staircase (below). Marvel at the limestone formations in front of you, and, as you listen to the roar of the underground waterfalls behind you, you'll be reminded of how this whole network was created by just the power of water. If you're brave enough, be sure to explore the Ghost Passage while you're down here too – it is just as mysterious as it sounds!

THE GLOWWORM CAVES

People have been visiting the caves for over 130 years. Not long after you set off on foot, you'll enter the Cathedral, a cave 150 ft. (45 m) below ground and 59 ft. (18 m) high. It's an incredible 30 million years old and has been shaped by the river running through it. Next, board a boat on the river and silently glide along the water into the soft blue light of the glowworm grotto. The ceiling is covered with thousands of these tiny creatures, lighting your way like stars in the night sky.

THE STATS

Location: Waitomo Glowworm Caves, North Island, New Zealand
Best time to visit: Oct.–Apr.
Depth: 150 ft. (45 m)
Temperature: 60.8–62.6°F (16–17°C)
Visiting time: 45 minutes
No. of caves: Hundreds, but just 10 are open for tours

DID YOU KNOW?

The glow emanating from a glowworm is created by chemicals in its body reacting with oxygen in the air. And it isn't actually a worm but is, in fact, a type of insect.

BLACK WATER TUBING

Head 262 ft. (80 m) down into the Ruakuri Cave, squeeze through a hole in the rock, and prepare to explore the underground river. You'll jump off waterfalls, float around bends, and follow a trail of twinkling lights with the glowworms as your guide. Those aged 12 and over can take on the adrenalin thrill of the Black Abyss trail (below). After abseiling down 115 ft. (35 m), you'll zip-line through pitch-black caves – the glowworms above being your only light source – tube your way along underground river rapids, and climb waterfalls. Luckily, experienced guides can help you along and will even provide sweet treats to keep you going until the end.

ARANUI CAVE

You won't find any water or glowworms here, but what you will discover are some of the most impressive stalactites and stalagmites you'll ever see (above). Limestone needles pointing up from the floor and hanging down from the ceiling sparkle in shades of pink, yellow, and white. They have taken thousands of years to form and are perfect examples of nature's artwork.

DISCOVERING THE GLOW

Local Māori people have always known about the caves' existence, but it wasn't until 1887 that Māori Chief Tane Tinorau and an English surveyor called Fred Mace first explored them. They took candles to light their way, but quickly realized they wouldn't need them as thousands of glowworms reflected their light onto the water below. This *Arachnocampa luminosa* species of glowworm can only be found in New Zealand. When glowing, they are actually in their larval stage, so should technically be called "glow maggots"!

MORE FUN!

The famous *Lord of the Rings* movie trilogy was filmed entirely in New Zealand. These fantasy films are as famous for their scenery as for their storyline. Just one hour from the Waitomo Caves, you can visit the famous Hobbiton movie set. In the stories, creatures called hobbits live here, playing a key role fighting to save their world from evil forces!

Bathe in warm, geothermal mud...

...AT HELL'S GATE

There's a whiff in the air in Rotorua, New Zealand, and it's not pleasant! But that smell means fun. The whole area is a geothermal hot spot, and the stink of eggy hydrogen sulfide in the air is a small price to pay for the experience you'll discover at Tikitere, also known as Hell's Gate. Describing itself as "Where the center of the Earth meets the sky," you'll find mud baths, steaming cliffs, mud volcanoes, and boiling hot pools.

IT'S HOT IN HERE

Begin your visit to Hell's Gate with a geothermal walk. This area puts on a show like nowhere else, as heat from deep below the Earth reaches the surface. Everywhere you look there are pools of super-heated water, including one that's 49 ft. (15 m) deep. The aptly named Devil's Bath (above) reaches 113°F (45°C) – but its neon water is acidic, so don't be tempted to jump in. The Sodom and Gomorrah pools have temperatures of up to 212°F (100°C) and shoot water almost 6.5 ft. (2 m) in the air, so unless you want to be scalded, it's these that you really need to be careful of!

STEAM SHOW

Steam rises all over Hell's Gate, creating an air of mystery. But between the clouds you'll see the Southern Hemisphere's largest hot-water waterfall. With a temperature of around 104°F (40°C), the Kakahi Falls were, understandably, a popular post-battle bathing spot for local Māori warriors. Farther along the walking trail, you'll find the Steaming Cliffs Pool (below). As the hottest pool in the reserve, the temperature of its water can reach up to 248°F (120°C) bubbling, spitting, and spurting as it boils vigorously.

DID YOU KNOW?

The Māori name for the area is Tikitere. The English name was suggested by the Irish playwright George Bernard Shaw when he visited the area in 1934.

MUD BATHS

Not all the pools here contain water – some are full of mud! The reserve has three types of mud in its pools: black, gray, and white. The Devil's Cauldron contains hot black mud that reaches up to 212°F (100°C). But the muddiest showstopper in the park is the Mud Volcano (right) – at 105 ft. (3.2 m) high, it erupts mud in the same way that a volcano erupts lava.

THE STATS

Location: Rotorua, North Island, New Zealand
Best time to visit: Nov.–Apr.
Hell's Gate Māori name: Tikitere
Formation: 10,000 years ago
Mud volcano eruptions: About every 6 weeks

FOOT SPA

At the end of your walk around the reserve, dip your feet in the silky, warm mud pool for a relaxing experience. If you'd like to take things further, you can head to the spa and fully submerge yourself!

UNDERGROUND OVEN

Māori people have long used the geothermal heat in the ground to their advantage – they cook with it by creating an oven-like space in the soil, known as a *hangi* (below). The free heat can be used to create tasty roast meats along with slow-cooked veggies. Within Hell's Gate there's also a historic cooking pool, with a temperature of 190°F (88°C). Food is placed into bags and lowered into the water to cook. Thankfully, the bags are waterproof so the murky water doesn't taint the taste of the food in any way!

MORE FUN!

A Māori tribe called Tūhourangi Ngāti Wāhiao lives in Rotorua and you can visit their traditional village, Whakarewarewa, to learn more about their culture and how they have used the geothermal landscape for bathing, cooking, and heating since their ancestors first settled there back in the 14th century. You will also be shown the famous *haka* war dance that many New Zealand sports teams perform before games.

Be brave and bungee jump...

...FROM THE KAWARAU GORGE SUSPENSION BRIDGE

Welcome to the adventure capital of the world, Queenstown in New Zealand. This town has the spirit of adventure in its DNA! Perhaps its most famous experience is the downright hair-raising activity of bungee-jumping. So it's time to pluck up some courage and test your limits as first you leap, then freefall, attached to planet Earth by just a single length of elastic.

TIME TO FLY

Bungee-jumping is all about height, flight, and speed... things many people are scared of. But overcoming a fear can be liberating, so you might want to give it a go! Once you've headed to your jump site, you'll meet your crew and harness up. Your heart will be pounding and your mind racing – do you choose to look over the ledge or not? Decide how you'd like to jump – leap, lean, fall backward, run, be pushed... there are lots of options. When you're ready, just go for it!

THE STATS

Location: Kawarau Gorge, South Island, New Zealand
Best time to visit: Sept.–Feb.
Height: 141 ft. (43 m)
First commercial bungee jump: 1988
Age limit: 10 years and over, and over 77 lb. (35 kg)

JUMP SITES

There are actually three bungee-jump sites to choose from in Queenstown. The original – Kawarau Bridge (left) – is where Queenstown's first bungee jump took place, and at 141 ft. (43 m) high, it's a great first jump option. The river Kawarau runs beneath, so there's the possibility of a "water touch," where you splash into the river before the cord springs you back up. Next, try the Ledge Bungee (above), a freestyle jump from 154 ft. (47 m) with great views across Queenstown. Those over the age of 13 can consider conquering the Nevis Bungee (right). At 440 ft. (134 m), it's the highest in New Zealand and offers 8.5 seconds of freefall. Yikes!

MAKING HISTORY

In the 1980s, New Zealander A. J. Hackett became intrigued by the ritual of jumping from heights and began to start work on developing a springy cord with his friend Henry van Asch. After endless tests, Hackett made headlines by bungee-jumping from the Eiffel Tower in June 1987! After an initial arrest, the police released him and the sport of bungee-jumping was officially born. The first commercial bungee jump then took place in 1988, from the Kawarau Bridge in New Zealand.

MORE FUN!

Based around beautiful Lake Wakatipu, Queenstown is as pretty as a picture. It's no wonder tourists love to visit. You can ride a gondola to the top of Bob's Peak to enjoy a 220-degree view of the area – it's up here you'll find the Ledge Bungee. You'll also be able to spot many of the other sports taking place in the area too: paragliding, quad biking, jet boating, zip-lining, and even brand-new activities like Fliteboarding – a cross between surfing and snowboarding – which takes place on Lake Wakatipu.

Swim alongside manta rays...

...IN THE YASAWA ISLANDS

Bula! Welcome to the island paradise in the South Pacific Ocean where you have the chance to encounter majestic manta rays. After a ferry ride from Fiji's main island, Viti Levu, you'll arrive at a group of islands called the Yasawas. Here, you'll find a shallow channel called Manta Ray Passage. It's one of the best places in the world to swim with these impressive creatures. Just listen for the locals sounding the lali drum indicating the manta rays' arrival in the passage, and within minutes you can go from lazing on the beach in a hammock to swimming in the water with these gentle giants.

GRACEFUL GIANTS

The rays that visit Manta Ray Passage are reef manta rays, and each one can be identified by its unique spotty markings. Another larger species, the giant oceanic manta, can be found farther out in deeper waters. The reef rays love to feed on the plankton that drifts through the Manta Ray Passage during high tide and, as filter feeders, all they need to do is swim with their mouths open to scoop up these tiny, tasty treats. Watch as they swim along by moving their fins like wings.

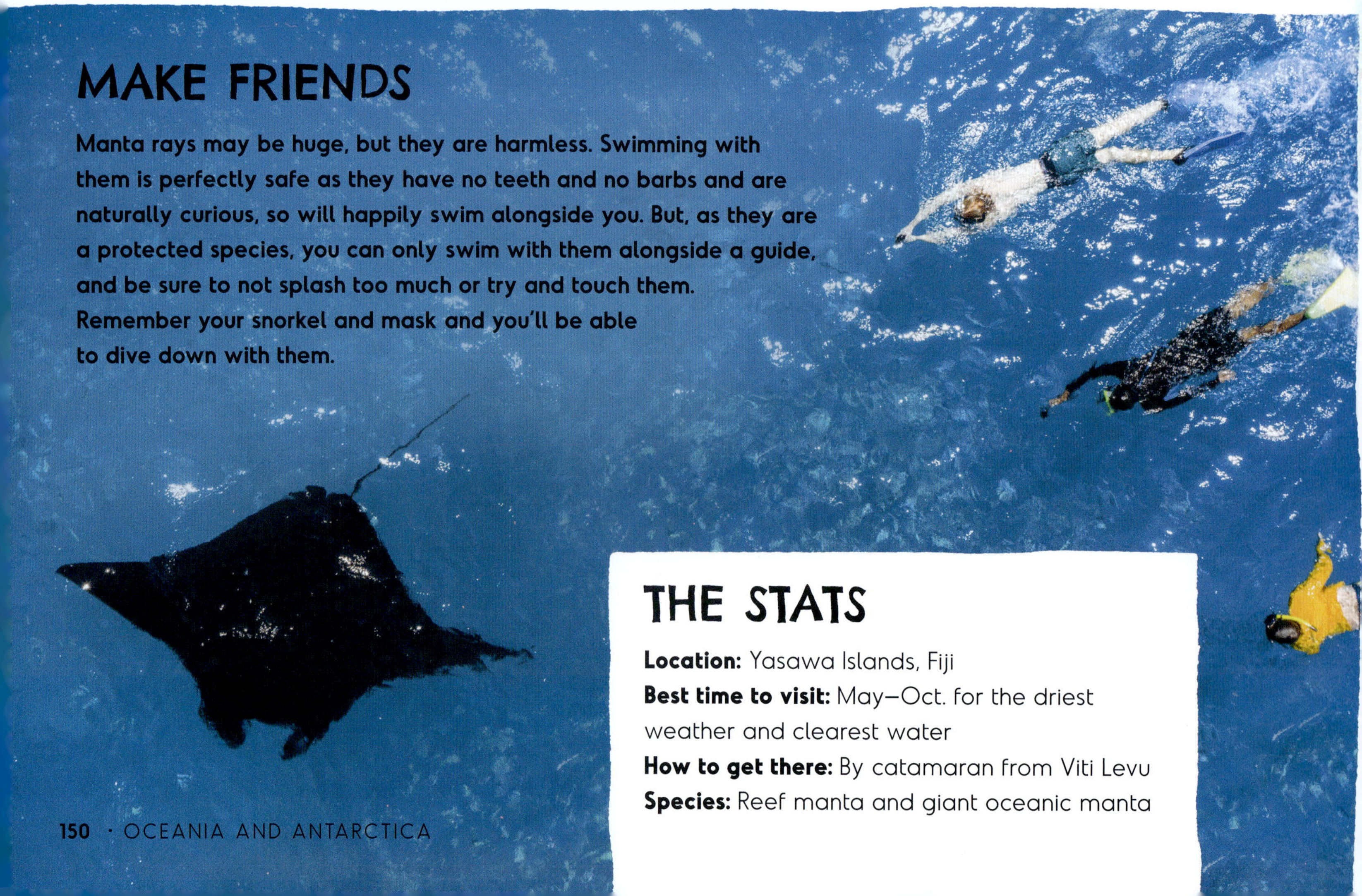

MAKE FRIENDS

Manta rays may be huge, but they are harmless. Swimming with them is perfectly safe as they have no teeth and no barbs and are naturally curious, so will happily swim alongside you. But, as they are a protected species, you can only swim with them alongside a guide, and be sure to not splash too much or try and touch them. Remember your snorkel and mask and you'll be able to dive down with them.

THE STATS

Location: Yasawa Islands, Fiji
Best time to visit: May–Oct. for the driest weather and clearest water
How to get there: By catamaran from Viti Levu
Species: Reef manta and giant oceanic manta

BODY IMAGE

The Spanish word *manta* means "blanket" or "cloak." The manta ray certainly looks like a blanket floating in the water, with its broad head, triangular fins, and flat body. Its eyes sit on the side of its head and it is believed to be one of the cleverest fish in the ocean, hunting by using both its sense of sight and smell.

DID YOU KNOW?

An amazing pink manta ray has been spotted off Lady Elliot Island in the Great Barrier Reef in Australia. The color is believed to be due to a mutation. Measuring 11 ft. (3.3 m) long, it's only been seen a handful of times.

CAREFUL CONSERVATION

The Manta Project Fiji is a group dedicated to conserving manta rays through research, education, and collaboration. Both species of the manta ray are listed as vulnerable and endangered, so it's really important that eco-tourism is promoted. This not only boosts employment in local communities but also keeps a cap on the number of people swimming with the rays so that disturbance is kept to a minimum.

MORE FUN!

With the beautiful ocean as a playground, any of Fiji's 330 islands offer a great day out. Explore the water by kayaking, snorkeling, or diving, and then relax on land where you can choose your level of luxury. To learn more about the culture, visit a local village and see a traditional *meke* dance (right). If you've visiting in July or August, you may also have the chance to witness a traditional firewalking ceremony at the Mariamma Hindu Temple.

Help save sea turtles...

...IN VANUATU

In the southern Pacific Ocean, a collection of 83 beautiful islands makes up the country of Vanuatu. The waters surrounding the islands are warm, providing the perfect conditions for five different species of turtles: the leatherback, the hawksbill, the loggerhead, the green, and the olive ridley. It's no wonder turtles are important to the people of Vanuatu – join in their conservation efforts and make a difference to the lives of these beautiful sea creatures.

IT'S ALL IN THE NAME

The hawksbill turtles that the conservation program protects get their name from their narrow, pointed beaks. They are perfectly shaped to pull out sponges from the coral reef to eat. By doing this, the turtles also help to improve the health of the reef, as sponges break down the coral. The turtles power themselves through the water using their long flippers.

THE STATS

Location: Moso Island, Vanuatu
Best time to visit: Apr.–Sept. for the dry season
Turtles released per year: Up to 300
Average water temperature: 80°F (27°C)
Number of turtle eggs: 60–200 per nest

HUMAN HELP

Visit the Tranquillity Island Resort in Vanuatu and get involved with their Hawksbill Turtle Conservation Program. Around 300 juvenile hawksbill turtles are raised here at any one time, from when they hatch until they are one year old. Time your visit right and you might get the chance to release a turtle back into the sea! Or you could opt to sponsor a baby hawksbill. You'll choose its name and receive an email whenever it is spotted swimming in the ocean after it has been set free.

DANGER AT EVERY TURN

Life is hard for turtles, right from the very start. As soon as they hatch they have to scuttle across the sand and into the sea for safety. But this is easier said than done, as predators such as birds and crabs see the turtles as an easy meal and many don't make it. Another reason sea turtles are endangered is because they have to travel huge distances to locate food and suitable nesting sites, and both are becoming harder and harder to find. They encounter danger in the ocean from humans, too – in the form of boats, nets, poaching, and pollution.

DID YOU KNOW?

Female sea turtles return to the same beach they hatched on to lay their own eggs.

MOUNT YASUR

Although much of life on Vanuatu is based around the sea, there's some interesting activity inland as well. Mount Yasur (left) is the country's constantly erupting volcano, and at just 1,181 ft. (360 m) high, it's very accessible – you can either walk up or go in a car. Once at the top, mail a letter at the "Volcano Post," the world's only mailbox on an active crater!

MORE FUN!

For underwater exploration head to Espiritu Santo, Vanuatu's main island. Here you'll find a World War II ship, SS *President Coolidge*, nestled among the reefs. But this isn't a typical warship – this vessel was a luxury cruise ship before she was used to carry troops in the war. On your dive you'll see guns, cannons, and other military equipment, along with beautifully decorated rooms, elaborate chandeliers, and mosaic tiles – plus plenty of reef fish that now call this wreck home.

Slice through snowy seas...

...TO SEE PENGUINS ON THE ICE

For many explorers, Antarctica represents the most extreme natural wilderness adventure on planet Earth. It's an untouched continent with otherworldly scenery and extraordinary wildlife. So layer up and set off for a trip over water, ice, and snow in search of Antarctica's most famous residents... the penguins.

PICK A PENGUIN

When most people think of Antarctica, it's penguins that spring to mind. There are 18 different species of penguins, but only two of them call Antarctica their true home: the emperor and the Adélie penguin. The chinstrap, gentoo, and macaroni penguins can be found breeding on the very edge of Antarctica where the conditions are a little less extreme, and it's here that you can kayak to see them. But if you really want to take on the iconic penguin experience, observing a colony of emperor penguins, then you'll need to hitch a ride on the ship's helicopter to Snow Hill Island and then walk across the ice to reach them.

WAY THROUGH THE WATER

To reach Antarctica you'll need to travel over water. So the first leg of your trip to find the penguins is aboard a huge icebreaker ship. These ships are specially shaped and strengthened to be able to break through the ice-covered waters, but their size means they need to stop some distance from the penguins so as not to disturb them. Your options are then to kayak to places like Cuverville Island to find gentoos, or hope for good weather so you can make the extra journey to the remote Snow Hill Island and the emperors.

SENSORY OVERLOAD

If you manage to reach the emperor colony, take a minute to let your senses settle. The smell of guano (penguin poop!) will hit you hard – it is somewhat pungent – but you'll quickly forget about it as the sound of parents and chicks chattering distracts you. And, as your eyes adjust to take in the number of penguins in front of you, you'll soon start focusing on the details of these majestic birds.

DID YOU KNOW?

Antarctica is a frozen continent, with 98 percent of it covered in ice. But in the summer months, some areas of the Antarctic peninsula defrost. Due to climate change, average temperatures here have risen by 5.4°F (3°C) over the past 50 years.

Spot the males incubating a single egg balanced on top of their feet. They keep it warm until it's ready to hatch.

Emperor penguins swim and dive for silverfish, krill, and squid. They typically eat about 4.5–6.6 lb. (2–3 kg) a day and, of course, bring back food for their chicks.

The colony works together to huddle for warmth. Each adult takes a turn on the outside where it's colder, and the chicks keep warm in the center.

THE STATS

Emperor penguin colony location: Snow Hill Island, Antarctica
Best time to visit: Oct.–Nov. to see the chicks
How to get there: Icebreaker ship, then helicopter flight, then walk across the ice
Number of penguins: Around 10,000
Time allowed on the island: 1 hour

MORE FUN!

Emperor penguins are pretty safe when they are on ice, as there are few predators – other animals are simply not built for the freezing conditions. However, they do need to watch out for the gulls that might try grabbing an egg or vulnerable chick. In the water there's more danger – sea lions, orca whales, and leopard seals (left) pose a threat to all penguins. These cold-water swimmers are iconic Antarctic animals, so try and spot them on your adventure, too. Birds to look out for include snow petrels, flocks of which like to sit on icebergs, and the wandering albatross, which has the longest wingspan of any bird.

Snowmobile to a research station...

...AND DISCOVER THE MYSTERIES OF ANTARCTICA

While penguins make up most of the population of Antarctica, up to 5,000 humans live here throughout the year conducting scientific research. While it may seem as though there's only ice to study, look a little harder and you'll discover there's more than meets the eye. Layer up your thermals and support the teams learning about our planet's coldest and least-visited continent.

EXTREME TRAINING

The United States' Antarctic Program (USAP) believes that the conditions on Antarctica closely parallel the conditions that humans will have to endure when on missions in space. This means they can use the bases to test both people and equipment for future missions to the Moon and Mars.

THE STATS

Location: Antarctica
Best time to visit: Nov.–Feb. for the summer tourist season
Number of research stations: 70
Summer population (Oct.–Feb.): Around 5,000
Winter population (Mar.–Sept.): Around 1,000
Coldest temperature: -128.2°F (-89°C) at Vostok Station

VISIT A RESEARCH STATION

The countries that operate the research stations in Antarctica arrange private transport for their staff, so there's no regular public transportation for tourists. But luckily you don't have to sign up as a crew member to visit. If you reserve a ticket on an Antarctic cruise, your itinerary will include a visit. Your ship will dock in the harbor, then a speedboat will take you ashore. If there's snow on the ground, station staff use snowmobiles to get around. These speedy vehicles have caterpillar tracks on the wheels for stability and skis at the front for steering.

SEARCHING FOR TREASURE

Despite the freezing temperatures, Antarctica is a hot spot for meteorites – and landing on an all-white background means they're usually easy to find. The United States funds a program that allows a team of up to 10 explorers to live for 5–7 weeks on the ice and use snowmobiles to scour the Transantarctic Mountains for meteorites, so they can be studied to help us learn more about our universe. Maybe you'll spot one on your travels?

SOUTH POLE STATION

On a plateau 9,301 ft. (2,835 m) above sea level, the Amundsen-Scott South Pole Station sits on adjustable stilts to prevent it becoming buried by snow. It's named after the Norwegian Roald Amundsen and the Briton Robert F. Scott, who led separate teams in the race to be the first to the South Pole in the early 1900s. Due to its position near the South Pole, it's the only inhabited place on Earth where there is continual daylight for five months (October to February) and continual darkness for five months (April to August). During the period of darkness, called polar night, temperatures can reach a bone-chilling -99°F (-73°CF) – understandably, it's not advisable to visit at this time of year!

DID YOU KNOW?

Many countries have set up stations in Antarctica. But the Antarctic Treaty signed in 1959 states that all observations from scientific investigations must be shared between teams.

MORE FUN!

Modern-day research stations have impressive design features and are often raised on stilts so snow can blow underneath them rather than build up around them. But when people first started exploring Antarctica, it wasn't like this. Anglo-Irish explorer Ernest Shackleton and his crew erected a simple wooden hut on their attempt to reach the South Pole over 100 years ago and it still stands today, frozen in time. Visit it and you'll see it even has cans of food sitting on the shelves from all those years ago!

INDEX

PICTURE CREDITS

The publisher would like to thank the following for their kind permission to reproduce their images
Key: l – left, r – right, t – top, b – bottom, c – center,

Shutterstock.com: anek.soowannaphoom Cover tl; proslgn Cover tr; Lianys Cover ct; Fotografie-Kuhlmann Cover cb; ArtDesign Illustration Cover cr; Rob Bruggeman Cover bl; michael nicolai Cover cb; David Ionut 1, 46; Elizaveta Galitckaia 4–5; Fotogrin 8 tr, 14 bl, 38 tr, 40 br; Ondrej Prosicky 8 cr, 44 br, 45 tl; Sakarin Sawasdinaka 19 tl, 125 tr; Byjeng 9 tr, 7 bl; NancyS 9 cr, 37 br; Kent Weakley9 bl, 24; Nick Fox 11 tr, 30 b; RuslanKphoto 11 tr, 28 br; worldswildlifewonders 11 bl, 37 cl; Lawrence Cruciana 11 br, 35 t; jabiru12–13; Daniela Constantinescu 13 tr; Unai Huizi Photography 13 cr; Elena Arrigo 13 br; Gimas 14–15 t; Matej Kastelic 14 br; Vacclav 15 tr; Sean Pavone 15 b; Idealphotographer 16 tr; TMP – An Instant of Time 16 bl; nito 17 tl; logoboom 17 c; Toms Auzins 17 br; Mia2you 18 t; Benny Marty 18 bl, 126–127; Galyna Andrushko 18 c; marion horan 18 cr; GJ-NYC 18 cb; imageBROKER.com 18 br; Anders Riishede 19 t; Susanne Pommer 19 cl; Nagel Photography 19 br; Pit Stock 20–21, 21 tr; Marcio Jose Bastos Silva 21 br; Shakhawat hossain007 22 tr; Wirestock Creators 22 cl; Junior Braz 22 cr; Christopher Chambers 23 t; Viaval Tours 23 tr; Luuk de Kok 23 cl; Mirigrina 23 br; Zhukova Valentyna 25 c, 25 br; Sergii Figurnyi 26–27 t; Lidiia Kozhevnikova 26 b; Songquan Deng 27 tr; Shawn.ccf 28–29 t; Chase Clausen 28 br, 29 tr, mimitza 29 tr; achinthamb 29 cr; SL-Photography 29 br; GTS Productions 30 tr; Suriel Ramzal 32 cl; Axolotl Photography 32 br; Kobby Dagan 33 tr; Anna Om 33 br; R Gombarik 34 tr, 34 b; igor_lavresh 35 bl; Daniel Huebner 35 br; Wollertz 36 tr; Simon Dannhauer 36 cl; Daniel Humberto Umana 36 br; Ondrej Prosicky 37 tl, 37 tr; buchpetzer 37 tc; Milan Zygmunt 37 c, 37 cr; fermorma 37 bl; Fluglinse 38 cr, 56 bl; Douglas Olivares 39 tr, 44 bl; Yongyut Kumsri 39 c, 47 cr; Nick Blamire-Brown 40 bl; Jose Hernandez Camera 51 41 tr; Kevin Oke Photo 41 cl; Danita Delimont 41 br; Celso Pupo 42 tr; Migel 42 br; A.Paes 43 tl; Altrendo Images 43 cr; Diego Grandi 43 br; FotoRequest 44 tr; Dr Morley Read 45 tr; Tarcisio Schnaider 45 br; salvatore ferri 47 tr; Satta Ektrakul 47 cl; Russell Johnson 47 bl; Eduardo Fernandez 48 br, 49 tr; ByDroneVideos 48-49 c; BonnieBC 49 cr; Marianna Ianovska 49 br; posztos 50–51, 51 tr; Vadim Petrakov Back Cover r, 52 bl; Douglas Olivares 53 tr; Pablo Rogat 54 bl; Maiaspops 54–55 c; Aleksandar Todorovic 55 bl; Skreidzeleu 56 cr; Fotografo de los Andes 56–57 t; Fabio Lamanna 57 cr; Scott Biales DitchTheMap 57 br; Imago 58 cr, 119 cl; David Osborn 58 br; Arthur Greenberg 59 tl; Amirkhans world 59 cr; Allen.G 59 bl; Pajor Pawel 60 cr, 73 br, 75 t; V. Belov 61 tl, 62–63 t, 62–63 b; Lizavetta 61 tr, 65 tr; Cristian M Balate 61 cr, 78–79 t; Rainer Albiez 61 bl, 82 bl; theycallmethewildrose 63 cr; Valmond 63 br; evgenii mitroshin 64 bl; Roman Babakin 64–65 t; Geoffrey Kuchera 65 br; Marcia Cobar 66–67 b; karenfoleyphotography 67 t; Sebw 67 tr; Viktorishy c; ESB Professional 68 cr; Stig Alenas 68 bl; RPBaiao 69 t; ElenaNoeva 69 cl; Frank Bach 69 br; Dignity100 70 cr; elRoce 70 b; Mistervlad cr; ChameleonsEye 71 bl; Xann Marketing 73 t; lexey Fedorenko 74 t; RichartPhotos 74 tr; chettarin 74 br; Tanasut Chindasuthi 75 cr; JudeAnd 75 bl; javarman 76 tr, 91 tl; Tatiana Popova 76 br; Vernerie Yann 77 tl; Alexandra Lande 77 cr; Byjeng 77 bl; holwichaikawee 78 bl; Calin Stan 79 br; Alfonso de Tomas 80 cr; BearFotos 80 b, 81 tr, 81 bl; FotoDuets 81 tl; Alexey Fedorenko 81 br; viewworld 81 tr; Alberto Masnovo 83 t; Nikiforov Alexander 83 cl; Luca Petralia Photography 83 bl; canadastock 84 b; Petr Pohudka 85 t; Piotr Poznan 86 cr, 97 c; Huang Zheng 87 tl, 92 c; AlexAnton 87 tr, 88; Matrishva Vyas 87 c, 91 cr; Sergey Uryadnikov 87 bl, 100 tr; Framalicious 87 br, 99 br; ahmedfawzyelaraby 89 tr; Barbara Brockhauser 89 cl; Orhan Cam 89 cr; zevana 89 b; Jane Rix 90–91 b; George Lamson 91 br; Curioso.Photography 92 br; KajzrPhotography 93 cr; Michail_Vorobyev 95 bl; Gudkov Andrey 96 cl; Anders Stoustrup 96 b; Henk Bogaard 97 tc, cr; Lenush 97 tr; Martin Mecnarowski 97 cl; Martin Harvey 97 br; David Havel 98 tr; Pozzo di Borgo Thomas 98–99 b; Dennis van de Water 99 tr; Jessica Seghatti 100 cr; Anya Newrcha 100 bl; EMBorque cl; Roger de la Harpe 101 c; Jake Keeton 101 bcl; francesco de marco 101 bc; Sergey Uryadnikov 101 bcr; Shawn Levin 101 br; Vadim Petrakov 102 cr; Anton_Ivanov 103 tr; Anna Dunlop 103 cl; Sean Heatley 103 cr; Great Stock 105 cr, 105 br; Sergii Figurnyi 106 tr, 109 tr; Ilona Ignatova 106 br, 110–111; Daniel Prudek 107 cl, 117 tl; Oscar Wistrand 107 tr, 122 bl; Patchanokk 107 bl, 126 bl; Guitar photographer 108 tr; AVVA Agency – Gozalov 109 cr; LiskaM 109 bl; Pandora Pictures 110 bl; marcobrivio.photography 111 tr; Stockbym 111 br; Shivani Hansraj Dhargave 112 bl; Kristin F. Ruhs 112–113; LightField Studios 113 cr; heFinalMiracle 113 br; Veranika848 114 cr; Anuradha Marwah 114–115; AlexAnton 115 br; Vixit 116, 116 bl, 117 cl; Daniel Prudek 117 tl; Saulius Damulevicius 117 cr; Kandarp 117 br; Pavel L Photo and Video 118 tr; DnDavis 118 br; Animalgraphy Back Cover cl, 119 bl; aphotostory 120 tr; Nateethep Ratanavipanon 120 bl; Alexander Studentschnig 120 bc; fotohunter 120 br; ABCDstock 121 tr; Esnal Julen 121 cl; yongyuth limpasute 121 br; BlueOrange Studio 122–123; Tetyana Dotsenko 123 br; KenSoftTH 124 cr; Daily Travel Photos 124 br; Indratauf 125 tl; lydiarei 125 cl; Benny Marty 125 cr; martinho Smart 125 br; Takashi Images 127 cr; dleffler 128 tr; Santiago Duarte 128–129 b; Steph Couvrette 129 cl; Lu Nhat Thuyen 129 cr; Sergey 402 130 bl, 154 bl; Andrei Armiagov 131 tr, 152 cr; paul white 131 cl, 142–143 t; Summit Art Creations 131 cr, 145 br; Manon van Os 132 bl; Tidewater Teddy 132–133; Christina Fink 133 cr; Tanya Puntti 134 tr, 135 cr; Islandjems – Jemma Craig 134 b; patileac 135 tl; Aaronejbull87 135 tcl; Ian Scott 135 tcr; fototrav 135 tr; Jenifer DeLemont 135 cl; StudioSmart 135 bcl; Imagine Earth Photography 135 bcr; Krofoto 135 bl; Frolova_Elena Back Cover tr; 135 br; sw_photo 136–137; Darren Tierney 137 tl; Erica Lorimer Images 137 tr; Bored Photography 137 br; Filipe Pires Castilhos 138 cr; Richie Chan 138; RugliG 139 tr; Jacksoo999 139 cl; Taras Vyshnya 139 cr; Natsicha Wetchasart 139 bl; Nico Faramaz 140–141; Amy Lutz 141 tr; Nigel Jarvis 141 br; K.A.Willis 142 br; Zioandre 143 cr; kwest 143 br; Lukas Bischoff Photograph 144 tr; Fotos593 145 tl; Filip Fuxa 146 tr; Evgeny Gorodetsky 146 b; Uwe Aranas 149 tl; mrmichaelangelo 149 br; Yann hubert 150 tr; Intertourist b; Shairaa 151 tr; NeHomo 151 cl; Nadezda Zavitaeva 151 br; StanislavBeloglazov Back cover bl, 153 cl; Nancy Pauwels 154 tr; Thijs van den Burg 154 br; Myeongho Seo 155 br

Alamy: Frank Gunn/The Canadian Press 10 cr, 28 cl; Robert Landau 16 cr; Helen Sessions b; Wolfgang Kaehler 31 tl; Olga Kolos 40 tr; Papilio 45 cr; John Warburton-Lee Photography 55 tr; Arctic Images 66 tr; Thomas Kyhn 68 br; CPA Media Pte Ltd 71 tl; Anwar Hussein 72 bl; horst friedrichs 84 cr; David Noton Photography 85 cr; SFM Titti Soldati 98 cr; Chris Howes/Wild Places Photography 103 bl; Greatstock 104–105 b; jbdodane 105 tl; Xinhua 107 br, 119 br; Auscape International Pty Ltd 143 tr; PA Images 147 bl; Boyloso 147 cr; imagebroker.com GmbH & Co. KG 152 bl; Media Drum World 153 b; Erik Lornie 155 c; Nature Picture Library 155 cr; NASA photo 157 tl; Vicki Beaver 157 c; Cavan Images 157 bl

Getty: sculpies Cover tr; Busakorn Pongparnit Cover cr; Westend61 2 t, 102; Markus Lange/robertharding 3, 148; Jan Sochor 11 c, 32–33 t; Stan Honda 20 bl; The Washington Post 25 tr; Vanderlei Almeida 43 tr; Paul Bereswill 48 bl; Jonathan Galione 51 br; Martin Harvey 52–53 c; DEA/C. Dani I. Jeske 53 cr; View Pictures 69 bl; Print Collector 72 tr; Max Mumby/Indigo 72 cr; Richard Pohle 72 bc; Leon Neal 73 bl; Thomas Barwick 93 b; guenterguni 94 cl; Roberto Moiola/Sysaworld 94–95; Rodger Bosch 101 tr; NicolasMcComber 104 cr; ozgurdonmaz 108–109 b; Jason Edwards 130 br, 156 br; Matteo Colombo 145 b; Tim Clayton – Corbis 145 cr; Mark Meredith 146 tr; Luke Mackenzie 149 tr; Jean-Francois Monier 149 cl; Alexis Rosenfeld 153 tr; Vicki Jauron, Babylon and Beyond Photography 155 cl

Other: Justin Foulkes/Lonely Planet Cover CR; Isabella Moore/Lonely Planet Cover br; Fábio Ribeiro da Silva/500px 39 br, 59 tl; ESO/G. Hüdepohl (atacamaphoto.com) 57 tr; Michael Heffernan/Lonely Planet 93 tl; NASA 156 t